Radiation Protection
in the Health Sciences

Second Edition

Radiation Protection
in the Health Sciences

Second Edition

Marilyn E. Noz
New York University School of Medicine, USA

Gerald Q. Maguire Jr.
Royal Institute of Technology, Sweden

World Scientific

NEW JERSEY · LONDON · SINGAPORE · BEIJING · SHANGHAI · HONG KONG · TAIPEI · CHENNAI

Published by

World Scientific Publishing Co. Pte. Ltd.

5 Toh Tuck Link, Singapore 596224

USA office: 27 Warren Street, Suite 401-402, Hackensack, NJ 07601

UK office: 57 Shelton Street, Covent Garden, London WC2H 9HE

British Library Cataloguing-in-Publication Data
A catalogue record for this book is available from the British Library.

RADIATION PROTECTION IN THE HEALTH SCIENCES — 2nd Edition
(With Problem Solutions Manual)

ISBN-13 978-981-270-596-9
ISBN-10 981-270-596-1
ISBN-13 978-981-270-597-6 (pbk)
ISBN-10 981-270-597-X (pbk)

Printed by FuIsland Offset Printing (S) Pte Ltd, Singapore

To our parents and to
Judge and Mrs. Edwin M. Clark

Preface

The discovery of X-radiation (W. C. Röntgen) and the subsequent discovery of natural radioactivity (A.H. Becquerel) and isolation of individual radioactive elements, particularly radium (M. S. Curie), were followed immediately by the employment of ionizing radiation in medical practice for both diagnostic and therapeutic purposes. At first the risks associated with the use of ionizing radiation were not fully appreciated. As ionizing radiation was more frequently and widely employed, these risks were uncovered and evaluated.

Since the first quarter of the previous century protection from ionizing radiation has been an important concern. As the medical uses of ionizing radiation have increased, the importance of understanding risk versus benefit has also increased. As the twentieth century progressed, advances in nuclear physics led to new applications which could expose the human population to ionizing radiation, e.g., nuclear power plants coupled with spent fuel rod disposal and plant decommissioning; and the manufacture, deployment, use, and disposal of nuclear weapons. Additionally, public consciousness concerning the potential effects of ionizing radiation on the human population increased due to such events as the accident at the Three Mile Island and Chernobyl nuclear power plants, by the dismantling of nuclear weapons, and by the development of medical diagnostic techniques such as ultrasound and nuclear magnetic resonance imaging which are alternatives to procedures involving radiation. This has resulted in the radiologist, health or nuclear physicist, and radiologic technologist frequently being called upon as experts in evaluating risk and protection measures, as well as having to answer questions concerning absorbed dose and its effect on the human body. In addition, as individuals these professionals have a personal interest, as well as professional requirement, to consider radiation safety in their work.

The purpose of this book is to help the reader understand and respect the recommendations relating to the safe use of ionizing radiation. It is intended for courses in an academic or training program for all those who may encounter ionizing radiation. It is directed primarily toward students preparing for a career as a radiologic technologist, medical physicist, health physicist, or radiologist. The practicing physicist, physician, technologist, or nurse will find it a source of material regarding radiation protection standards and techniques. The tables and graphs shown throughout the book are only illustrative and are only to be used for study. For professional practice, the latest primary sources for data should be utilized.

This book begins with a description of the fundamental processes by which ionizing radiation is produced and interacts with matter. It details the operation of radiation detection instruments and their use as survey and personnel monitors; includes the basic biologic effects of ionizing radiation along with the current units, regulations, and recommendations which govern exposure of radiation workers and the general public to ionization radiation; reviews the recommendations regarding the use of medical X-ray generators and radionuclides, and discusses good working habits associated with the use of such sources. This book also delineates the general and specific means used for protection against external ionizing radiation and explicates how to calculate absorbed dose from both internally deposited radionuclides and external sources of radiation. Each chapter contains review questions as well as problems. SI units are used extensively throughout the text as it is necessary to know and understand them. The appendices contain a discussion of units and logarithms. An extensive glossary is provided as well as and a bibliography which includes the relevant publications of national and international organizations. Answers to the problems are provided at the end of the book.

As with any book, the contents reflect criticisms and suggestions from numerous people. We would like to thank in particular the following: the many instructors and students who have used the two editions of the former book entitled *Radiation Protection in the Radiologic and Health Sciences* and the first edition of this book, our colleagues in the Department of Radiology at New York University, the Institute of Medical Radiation Physics at the Karolinska Institute and Stockholm University, and the Department of Communication Systems at the Royal Institute of Technology (KTH) in Stockholm, Sweden.

Marilyn E. Noz
Gerald Q. Maguire, Jr.

December 2006

Contents

Chapter 1

FUNDAMENTAL CONCEPTS

The phenomenon named "X-rays" by their discoverer, the physicist Wilhelm Conrad Röntgen, was the beginning of the what are today the fields of Diagnostic Radiology and Radiation Oncology. On November 8, 1895 Röntgen observed that electrons (then called cathode rays) accelerated through an evacuated tube generated rays which penetrated a light-tight enclosure and caused a barium platinocyanide screen to fluoresce. He spent the next weeks thoroughly investigating this phenomenon and published his first of three papers about it on December 28, 1895. Early in his investigations, he exposed a photographic plate to X-rays and found that different materials stop X-rays in different ways. His photograph of his wife's hand showed the bones and soft tissue as well as her ring had different degrees of permeability.[1] Although other investigators had observed X-rays, it was Röntgen who understood the full implications of their use and worked to exploit their potential for the good of humanity.

The discovery of X-rays was followed closely in 1896 by the confirmation of natural radioactivity by Antoine Henri Becquerel and the isolation of the radium by Marie Sklodowska Curie and her husband, Pierre Curie[2]. In the early part of the twentieth century, it was not at first recognized that either X-rays or radioactivity could be harmful as well as beneficial. This was first explicated carefully in the 1930's and the study of biological effects of ionizing radiation was undertaken in several countries. Both the consequences of the low (and at that time, not so low) amounts of radiation received by radiologists and others in their daily work, as well as the natural background radiation to which we are all exposed was extensively investigated.

1. Wilhelm Conrad Röntgen (1845-1923) refused to take out patents on his discovery so that the world might have it. At the time of his discovery, Röntgen was professor of physics at the University of Würzburg, Germany. He was awarded an honorary Doctorate of Medicine by Würzburg. He contributed his Nobel Prize (1901) money to the University of Würzburg. He was made Professor of Physics at the University of Munich in 1900 where he remained for the rest of his life. The old unit of exposure, roentgen, was named for him.

2. (Antoine) Henri Becquerel (1852-1908), a French physicist, discovered gamma-ray photons in 1896. Madame Marie Sklodowska Curie (1867-1934), working in France, though of Polish birth, and her husband Pierre Curie (1850-1906), isolated radioactive nuclei, significantly, radium. Becquerel and the Curies received the Nobel Prize in physics in 1903 for their work in radioactivity. Marie Curie also received the Nobel Prize in chemistry in 1911 for her work with radioactivity. The old, curie, and new unit, becquerel, of radioactivity, are named after these scientists who did pioneering work in the field of radioactivity.

Since then, the United Nations Scientific Committee on the Effects of Atomic Radiation[3] (UNSCEAR) has stated that medical use of radiation is the largest single source of man-made radiation exposure to the world population and that this source is increasing, both because more high dose procedures such as computed tomography (CT) are being used and because the use of diagnostic and therapeutic radiological procedures increases as more countries join the developing world. Medical use includes diagnostic radiology, radiotherapy, nuclear medicine, and interventional radiology. Additionally, radiation exposure occurs as a result of various occupations. Industries such as materials testing and nuclear power, as well as research which uses radiation or radioactive substances may be a source of exposure to workers. Radiation exposure to higher than natural levels of cosmic radiation is also experienced by airline passengers and crews and by astronauts.

The United States Food and Drug Administration (FDA) estimates that at least 130 million Americans have one or more diagnostic X-ray or nuclear medicine examinations annually. The FDA contends that these examinations account for more than 90% of the total man made radiation exposure to the U.S. population. While medical procedures account for only 15% of the total radiation exposure experienced by mankind, care should be taken to reduce it to as low a level as possible. The International Commission on Radiological Protection (ICRP) has emphasized [ICRP, 1982]

"the importance of including adequate training on radiation protection in the general education and training of individuals entering the medical and associated professions, since all those who enter these professions may be involved in prescribing procedures involving exposure to ionizing radiation. More thorough training in radiation protection is required by those planning to enter the field of radiology and by scientists and technicians assisting in the medical uses of radiation."

The United States National Council on Radiation Protection and Measurements (NCRP) showed that public radiation exposure from nuclear power plants is indistinguishable from background radiation [NCRP, 1987]. This report also states that occupational radiation exposure (less than 0.3%) coupled with radiation exposure from nuclear fallout (less than 0.3%), the nuclear fuel cycle (less than 0.1%) and other miscellaneous sources (0.1%) comprise less than 1% of the total radiation exposure to the population. Nevertheless, training should be extended to all individuals working with radiation, so that the exposure each one experiences may be kept **as low as possible**.

1.1 Need for Radiation Protection

One aim of radiation protection is to encourage and enable the radiation worker to limit his/her exposure to potentially harmful radiation. A second and equally important aim is to limit unnecessary exposure of others (e.g., patients and the general public) which might result from poor working habits.

---

3. See bibliography at end for references quoted in this and each succeeding chapter.

Various forms of radiation have always been present in mankind's natural environment. This radiation is known as background radiation and is generally divided into two categories: 1) radiation from specific forms of chemical elements that exist naturally in the earth's crust, the human body, and in materials about us; 2) radiation carried to the earth from outside the earth's atmosphere, commonly referred to as cosmic radiation.

In the twentieth century mankind learned how to produce radiation from various sources, and its uses have consistently increased. For example, we now use radiation techniques to produce nuclear energy for power, to sterilize both food and equipment (such as surgical and medical supplies), to produce new chemical polymers, to detect faults and flaws in metal castings and welds, and to aid research in areas such as medicine, biology, and material science. The most devastating use we have found for radiation is, of course, nuclear weapons, while the most widespread application of radiation has been in the field of medicine both for diagnosis and radiation oncology.

As is well-known, early radiation workers, such as technologists, radiologists, surgeons, and physicists, suffered severe radiation injury because they did *not* appreciate the extent to which radiation can cause injury. Many studies have been undertaken to further our understanding of the biologic effects of radiation and to establish acceptable limits of effective/equivalent dose (see Chapter 4 for specific details and examples). Some limited legislation has been enacted with this in mind, and regulatory bodies and licensing mechanisms have been established to set limits for radiation exposure as well as to delineate requirements for training programs for radiation workers. Their aim is to produce a uniformity of accepted practice in working with radiation.

The two general classifications of radiation are ionizing[4] and nonionizing. Ionizing radiation is capable of ionizing an atom. Ionization occurs when one of the orbital electrons of an atom has been completely removed. The residual atom, which is positively charged, is called a positive ion, or cation; the freed electron is known as a negative ion, or anion.

Radiation which creates ions that are then capable of disrupting life processes is known as ionizing radiation. In contrast, nonionizing radiation, such as ultraviolet and microwave radiation, lacks the ability to create ions. Nevertheless, nonionizing radiation can adversely affect human health. This book, however, is limited to a discussion of protection from ionizing radiation.

1.2 Ionizing Radiation and Injury

Transfer of energy to the human body may be beneficial, as is the case with food, or injurious, as is the case with a speeding bullet. Ionizing radiation always transfers energy to any material with which it interacts. The energy it deposits in living tissue causes disruption of the atomic structure, and when the atoms thus affected are essential for normal functioning a cell can be permanently damaged or killed. When ionizing radiation imparts

---

4. See Glossary for definition of this term and of many others.

energy to living tissue, damage is done: the larger the amount of energy deposited, the more extensive is the damage. Sometimes, for example, in radiation treatment for cancer, this damage can be *both* beneficial and injurious; it kills the tumorous cells, but it may also kill healthy cells. It is the transferring or depositing of energy in living tissue that is significant in the production of injury by ionizing radiation. As a result all measurements and calculations to evaluate the hazard from ionizing radiation have, as their initial object, the determination of the energy imparted by this radiation to the region of interest.

It has been epidemiologically demonstrated that biological damage is produced by ionizing radiation. Since there is no practical alternative, the ICRP strongly recommends that the relationship between radiation exposure and biological effect be treated cumulatively. An extensive study of the effects of low level irradiation over relatively long periods of time was commissioned by the United States National Research Council and carried out by the United States National Academy of Sciences. A committee was appointed to study the Biological Effects of Ionizing Radiation (BEIR). The results of the latest committee, BEIR VII, were published in 2006.

Three schools of thought exist regarding the effects of low levels of radiation exposure. In the first BEIR report, a linear hypotheses was proposed, meaning that the effect was exactly proportional to the amount of exposure (curve A in Figure 1.1). Later, some researchers felt that a low-limit cutoff for harmful effects may exist (curve B of the same figure). This cutoff is known as the threshold effect. Most recently, some research groups have suggested that effects at low levels of radiation exposure may be more severe than those at higher values (curve C). This controversy is difficult to settle. The BEIR III committee report presents arguments for and against each of the alternatives, the BEIR V committee report continues the discussion, and the BEIR VII report concludes that there is no biological evidence to support rejecting the linear, no threshold model.

Because the body cannot sense exposure to radiation directly, except at levels that are invariably lethal, it cannot provide a defense. Hence, it is important to be able to anticipate radiation problems through calculation and analysis. It is equally important to properly use those radiation instruments designed to monitor the emissions from radiation sources.

1.3 The Structure of Matter

Atoms are composed of a dense nucleus which has a radius of about 10^{-14} m (10^{-12} cm) with electrons orbiting about the nucleus at fixed energy levels resulting in an overall atomic radius of about 10^{-10} m (10^{-8} cm). Hence the atom is mostly empty space. Atoms differ from one another in the number and arrangement of their basic constituents: the nucleus is composed of protons (which have a unit positive charge, i.e., $1.6 \cdot 10^{-19}$ C), and neutrons (which have no charge); electrons (which have a unit negative charge) form the remaining atomic component. It is the number of protons in the nucleus which determines the chemical element and the number of electrons which determines the chemical

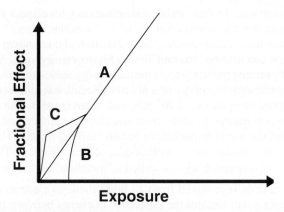

Figure 1.1: Relationship between radiation effects and exposure: Curve A represents the original linear relationship hypothesis. Curve B shows the possible cutoff of effects at low exposure levels (threshold effect). Curve C shows the possible enchancement of these effects at low exposure levels.

properties of an atom. When an atom is electrically neutral, the number of electrons is equal to the number of protons. The number of protons in the nucleus is symbolized by "Z" (the atomic number) and the sum of the number of protons plus the number of the neutron neutrons is symbolized by "A" (the atomic weight or mass). However, if one sums the mass of the protons ($1.6726 \cdot 10^{-27}$ kg), neutrons ($1.6747 \cdot 10^{-27}$ kg), and electrons ($9.1091 \cdot 10^{-31}$ kg, note that this is about 1/1800 the mass of the proton or neutron) composing an atom, the sum would be slightly more than the actual weight of the atom. This difference in mass is termed the "mass defect" and represents the binding energy of the nucleus. One often measures the atomic mass in terms of atomic mass units (AMU). One AMU is 931.2 MeV and is equivalent to 1/12 of the atomic mass of ^{12}C.

1.3.1 Atomic Electrons

Electrons are bound to the atomic nucleus by the Coulomb force (unlike charges attract, like charges repulse; the magnitude of the force is proportional to the amount of charge and inversely proportional to the square of the distance between the charges). Hence, electrons closer to the nucleus are more tightly bound to it, experiencing a larger force. Equally, the more protons in a nucleus, the more tightly the electrons closest to it are bound. The electrons in the atom normally occupy fixed energy levels usually referred to as "shells". The first allowed shell is the "K" shell, which can contain two electrons. The next is the "L" shell which can contain eight electrons, the next is the "M" shell which can contain eighteen electrons, etc. However, there can be no more that eight electrons in an unfilled outer shell of an atom and so it happens that after an element contains enough protons in the nucleus to require more than eight electrons in the "M" shell, but not a full eighteen, the electrons start to fill the next available shell (in this case the "N" shell, until eight is reached, etc.) This gives chemical elements, such as metals (which are commonly called "transition elements"), their particular properties.

When electrons are in their stable configuration with respect to the nucleus, they have associated with them an energy known as the "binding energy". This is the energy necessary to remove the electron entirely from the atom and is known as the "ionization" energy. An electron can also be "excited" into a higher energy state. When this happens, the electron usually returns promptly to its normal energy state, radiating the excess energy in the form of electromagnetic energy or a photon. (Recall that all electromagnetic energy travels with the speed of light, i.e., $3 \cdot 10^8$ m/s, and is governed by the equation $\nu\lambda = c$. Electromagnetic waves transport energy from one place to another, and this energy travels, or is transported by, the waves in packets or bundles termed quanta. These energy packets give electromagnetic waves their particle-like characteristics. The particle which characterizes an electromagnetic wave is called a "quantum" or "photon".)

When an electron is removed from an inner shell, an electron from an outer shell generally takes it place and radiates the difference in energy between the two shells in the form of a high energy photon. Since the energies involved are absolutely characteristic of the nuclear configuration of the atom, these photons are known as "characteristic X-rays" or "characteristic X-ray photons". An alternative process to the expulsion of a characteristic X-ray, is for a second atomic electron to be expelled. This electron is known as an "Auger" electron and it should be noted that now the atom has become doubly ionized. Note that electrons continue to fill in "lower" (i.e., closer to the nucleus) energy states until the missing electrons (or "holes") are in the "valence" or outermost occupied shell. Figure 1.2 shows a schematic representation of an atom, with an "M" to "K" transition accompanied by production of a characteristic X-ray. (Note: an Auger electron could have been produced instead of the characteristic X-ray.)

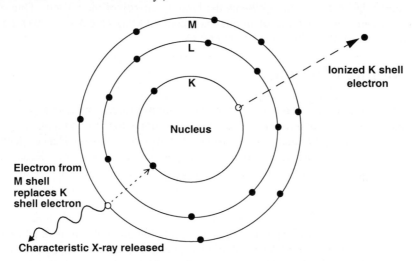

Figure 1.2: A schematic representation of atomic energy levels. Note that there are only eight electrons in the "M" shell. An electron has been removed from the "K" shell and an "M" shell electron fills in the vacancy and releases a characteristic X-ray.

1.3.2 The Nucleus

The nucleus contains protons (which all have a positive charge) and neutrons in a very small volume. Since like charges repel, the nucleus must be held together by a force greater than the Coulomb force. This force is known as the "strong" force and its exact nature is not yet well understood. One model of the nucleus postulates that the protons and neutrons exist in "shells" similar to electron shells. If one examines the periodic table, one finds that the number of neutrons increases disproportionately to the number of protons in the nucleus, thus suggesting that the neutrons help to keep the nucleus together as the proton number increases. However, after a certain number of protons, no nucleus is stable, i.e., all nuclei spontaneously undergo transitions such as nuclear fission, or far more frequently, radioactive decay (this will be discussed in detail in the next section).

If the number of protons in the nucleus is odd, the nucleus has a property known as nuclear magnetic resonance. Similar properties result if the number of neutrons is odd. If both are odd, these properties are magnified. However, if both are even, these magnetic properties do not exist. The nuclear constituents are known as "nucleons" and different nuclei are referred to as "nuclides". Nuclides which are related in various ways are referred to by various names as shown in Table 1.1.

Table 1.1: Nuclear Nomenclature

Name	Atomic Number - Z	Neutron Number - N	Atomic Mass Number - A
isotope	same	different	different
isotone	different	same	different
isobar	different	different	same
isomer	same	same	same

From the table it is evident that isomers differ from each other only in energy. This leads us to the idea that the nucleus itself has distinct and fixed energy levels associated with it which are absolutely characteristic of the nuclide. When a nuclide is in one of its excited states, it usually returns to the ground state with the expulsion of one or more gamma-ray (γ-ray) photons. The energy of these photons reflects the difference in energy between the allowed energy states of the nuclide. If an excited nuclide does not return to its normal ground state in a measurable ($\sim 10^{-11}$ s) period of time, it is said to be in a metastable state. Some nuclides, like ^{99m}Tc, remain in these metastable states for hours. As in the case of atomic electrons, an alternative process to the production of a gamma-ray photon is the production of an electron, in this case known as an "internal conversion electron".

The only difference between a gamma-ray photon and a characteristic X-ray photon is in their origin - the first results from a nuclear interaction and the second from an atomic interaction involving inner shell electrons. It is also important to understand that photons in the energy range denoted by these names are capable of ionizing atomic electrons. Interactions which involve outer shell, or valence electrons, usually result in photons having energies in the visible light region that are not capable of ionizing atomic electrons.

It was Albert Einstein who developed the equation relating the mass of a particle to energy, or the energy of a wave, to its mass equivalent.The equation is $E = mc^2$ where "c" is the velocity of light as mentioned above. This equation can be used to find the energy equivalent for the mass of a stationary (i.e., non-moving) electron: 0.511 MeV (see Glossary). Hence we can use the terms "mass" and "energy" interchangeably.

1.4 Radioactive or Nuclear Decay

Nuclides which undergo natural or induced transformation or decay of the atomic nucleus are said to be radioactive. A nuclide undergoes decay by one or more of the decay modes known as alpha, beta, and gamma or isomeric decay.

1.4.1 Alpha Decay

The first radioactive decay method is known as "alpha decay". When a nuclide expels an alpha particle (this particle consists of two protons and two neutrons) the energy of the particle is characteristic of the nuclide from which it came. Since the original or "parent" nuclide has lost two protons, the "daughter" or "offspring" nuclide is a different chemical element. The Z number has been reduced by two and also the neutron number, hence the A number of the new nuclide differs from that of the original nuclide by four. Since it is rare that the daughter nuclide is produced in its ground or lowest energy state, alpha decay is generally accompanied by gamma-ray emission and internal conversion electrons.

1.4.2 Beta Decay

The next method by which a nuclide may undergo radioactive decay, is known as beta decay. This type of decay mechanism comes in three varieties known as beta minus (β-) or negatron decay, beta plus (β+) or positron decay, and electron capture.

In negatron decay, a neutron in the nucleus decays into a proton (which remains in the nucleus) and an ordinary electron which is expelled from the nucleus. However, to obey all the conservation laws, vis., energy, angular and linear momentum, and lepton number (see Glossary), a second particle, known as an anti-neutrino, is also expelled from the nucleus. Hence:

$$N \rightarrow P + e + \bar{\nu} \qquad (1.1)$$

The total available energy (the difference in mass defect between the parent and daughter nuclides among other things) is shared between the electron and anti-neutrino in a random manner, so that the total energy for the decay is fixed ($E_{\beta max}$), but the specific energy that a particular beta minus particle possesses is random until the maximum is reached. However, the average energy of the beta minus particle is $\frac{1}{3} E_{\beta max}$ (see Figure 1.3).

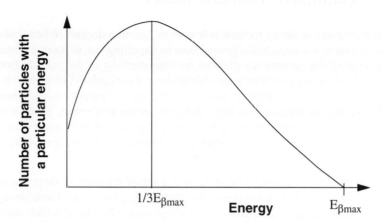

Figure 1.3: Energy spectrum of β particles. There is a maximum energy associated with any particular decaying nuclide, however, the total energy available is shared between the β particle and the (anti-)neutrino.

Since the daughter nuclide has a Z number which is one greater than the parent nuclide (A remains the same), it is a different chemical element.

In positron decay, a proton in the nucleus decays into a neutron (which remains in the nucleus) and a positive electron or positron along with a (real, not anti-) neutrino (for the same reasons as before) are expelled from the nucleus. Hence:

$$P = N + e^+ + \nu \qquad (1.2)$$

Again the total energy available is divided between the positron and the neutrino, so the energy of the positron is not fixed. There is a maximum and average energy as defined previously (see Figure 1.3). In this case, the neutron mass is greater than the proton mass, so energy must be put into the system in order for the decay to occur. Thus the difference in mass defect between the parent and daughter nuclide must be at least 1.02 McV for positron decay to occur.

An alternative to positron decay is electron capture. In this decay mode, an orbital electron is captured by the nucleus and combined with a proton to form a neutron. Only a neutrino is expelled from the nucleus. However, since an orbital electron is captured, this mode of decay is always accompanied by characteristic X-rays and Auger electrons. This mode of decay can always occur and in fact when positron decay is energetically possible, it is always accompanied by electron capture. In both positron decay and electron capture, the Z number of the daughter nuclide is reduced by one from the parent nuclide, so these are always different chemical elements. (The A numbers remain the same.) In all three modes of beta decay, the daughter nuclide is almost always produced in an excited nuclear state. Hence, beta decay is almost always accompanied by gamma-ray photons and internal conversion electrons.

1.4.3 Gamma or Isomeric Decay

The last radioactive decay method is known as "gamma decay". When a nuclide exists in an excited state, it returns to the ground state by expelling one or more gamma-ray photons. The energy of the gamma ray photons are characteristic of the nuclide from which they came. As mentioned previously, an alternative to the expulsion of a gamma-ray photon is an internal conversion electron. Since the original or "parent" nuclide has not lost any nucleons from the nucleus it is not a different chemical element. Since it is rare that the daughter nuclide in either alpha or beta decay is produced in its ground or lowest energy state, these decay methods are generally accompanied by the emission of gamma-ray photons and internal conversion electrons.

Each time a radionuclide undergoes a transformation or decay, a new nuclide is produced and is usually accompanied by other decay product(s). These decay products are usually, but not always, an alpha or and/or a photon. The rate at which radionuclides in a given sample undergo transformation, and, consequently, the rate of emission of the decay product(s), is directly proportional to the number of radioactive nuclides contained therein. Thus as the number of radioactive nuclides in the sample or source decreases because of the radioactive decay taking place, the rate of emission of the decay product(s) also decreases. The rate of nuclear transformation or decay is known as the activity of the source. This leads to an exponential decay curve, as shown in Figure 1.4. The time it takes for the activity to decrease to one half of its initial value is known as the half-life of the radionuclide. This will be more fully discussed in Chapter 11.

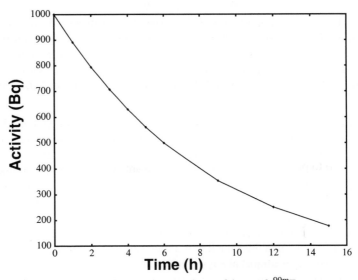

Figure 1.4: Exponential decay curve showing the natural rate of decay of ^{99m}Tc.

Every chemical element has associated with it an atom that has a fixed number of protons in its nucleus and the same number of electrons orbiting about the nucleus. However, the number of neutrons present in the nucleus may vary (and often does). One example is oxygen. The most common form of naturally occurring oxygen, which has eight protons and eight neutrons in the nucleus, also has stable isotopic forms which have nine or ten neutrons in the nucleus. At the same time, oxygen has another form (seven neutrons in the nucleus, $^{15}_{8}O$) which decays spontaneously, emitting a positive electron, or positron. When this nuclide decays, either by emitting excess energy in the form of a photon or by emitting a particle, or both, the nuclide is referred to as a radionuclide and is said to be radioactive. For example, most organic substances contain a very slowly decaying form of carbon (carbon-14, $^{14}_{6}C$). Due to this slow decay the ratio of ^{14}C to ^{12}C is used for dating organic materials.

Some radionuclides exist in nature; common examples are uranium-238 (^{238}U) and potassium-40 (^{40}K). Other radioactive nuclides are produced in special machines. For example, tellurium-124 (^{124}Te) can be used as the target material in a cyclotron. When this nuclide absorbs a high-energy proton produced in the cyclotron, it forms iodine-123 (^{123}I) plus two neutrons. ^{123}I then decays spontaneously emitting photons and electrons. The various radioactive decay processes are too extensive to be discussed here [Lederer, 1978].

In the SI system of units (see Appendix A) the unit of radioactivity is the becquerel (Bq). It is equal to one (disintegration, or decay) per second. In the former system of measurement, the unit of radioactivity was the curie (Ci), defined as $3.7 \cdot 10^{10}$ (disintegrations) per second which represents the amount of activity in one gram of radium-226. Since the curie represents a very large number of disintegrations, it was common to express activities in units of mCi (one thousandth of a curie), or μCi (one millionth of a curie). Note that 1 Ci is 37 GBq, 1 mCi is 37 MBq, and 1 μCi is 37 kBq. Appendix A gives an overview of the history of unit standardization and its current status.

1.5 Direct and Indirect Ionization

Ionizing radiation consists of charged particles, which are capable of ionizing directly, and/ or of particles that, though uncharged, are capable of producing charged particles through a secondary mechanism. Electrically charged particles with sufficient kinetic energy (energy of motion) to produce ionization by collision are called directly ionizing particles. Uncharged particles, which produce ionization only through secondary mechanisms, are known as indirectly ionizing particles.

1.5.1 General Interaction Mechanisms

An atom can be ionized in several ways. The method involving ionizing radiation usually occurs as follows: An electrically charged particle, which has sufficient energy (by virtue of its motion) to release an electron from its orbit, undergoes an inelastic collision with an orbital electron, transferring to it energy at least equal to that of the binding energy of that electron. This transferred energy enables the release of the electron from its orbit. Rarely does the initial charged particle, except in the case of fission fragments, interact with the atomic nucleus.

In general, charged particles are divided into three categories: 1) light particles such as electrons, positrons, and mu mesons, 2) heavy particles such as protons, pi mesons, and alpha particles, and 3) very heavy nuclear fission fragments. In the case of the first two categories, a charged particle, when it passes through matter, interacts mainly with atomic electrons. Charged particles that are light, like electrons undergo interactions within matter that are similar in the nature to two ping-pong balls colliding. Thus the path of light-charged particles through matter is not straight. Figure 1.5 is an attempt to illustrate this. As the incoming particle travels through matter, it loses energy, slows down and, therefore, the distance it travels between collisions is shorter. The incoming light particle transfers some of its energy to an atomic electron, thus giving it energy. Since both particles are approximately the same size, both have their direction of motion changed at least by a small amount. When the particle has slowed down sufficiently, it is trapped in orbit by an atomic nucleus. The ionization which light-charged particles produce is more concentrated toward the end of the path than toward the beginning. If the interacting material has a high atomic number, or if the energy of the charged particle is low, the probability that the light particle would be deflected from its path by more than 90° is increased.

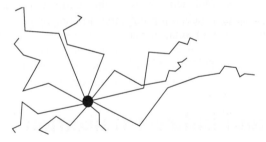

Figure 1.5: Tortuous paths that might be expected from a source of charged particles which are light, such as electrons, when they interact with matter.

However, in the case of a heavy-charged particle, the interaction with the atomic electron, which occurs in the form of collision, can be thought of more in terms of a bowling ball interacting with a ping-pong ball. The bowling ball transfers some of its energy to the ping-pong ball, but it is not noticeably slowed down or deflected from its path. The ping-pong ball on the other hand, because it has a small mass, usually moves quite rapidly away from the interaction site. After the collision, the heavy-charged particle has lost some energy, so the distance it travels before interacting again is shortened. This process, of course, repeats itself. Eventually, the heavy-charged particle has lost so much energy that it moves a very short distance between interactions, hence the volume of the matter surrounding the particle at this point is very heavily ionized. When a stable particle, such as a proton, has lost all or most of its energy, it gathers enough electrons around it to neutralize its charge and effectively stops. If one were to picture the amount of ionization that occurs along the path of such a particle, one would find that it is very sparse at first, gradually increases, and finally, becomes quite intense. The charged particle travels in a relatively straight line, and the distance it travels in an absorbing material is a function of its original energy when it enters the material as well as the electron density (i.e., the number of electrons per unit volume) of the material. Figure 1.6 is an illustration of the manner in which a charged particle which is heavy, typically loses energy. For example, protons having an energy of 5 MeV are all stopped by 0.4 m of air, whereas those protons having energies of 10 MeV require 1.0 m of air measured at 15°C and standard pressure (one atmosphere or 101.3 kPa) before stopping.

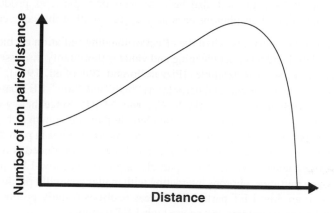

Figure 1.6: The specific energy lost by the interaction of a beam of charged particles which are heavy as it passes through matter. All particles are assumed to have the same initial energy.

Fission fragments, which are produced as a result of neutron-induced or spontaneous fission of heavy nuclei, contain nuclides of mass numbers from approximately seventy-two to one hundred sixty and start out having a positive charge of twenty to twenty-two. The initial energy possessed by fission fragments is also high, 65 to 97 MeV. The interaction of a fission fragment differs from that of other heavy-charged particles in two important aspects. First, it interacts not only with atomic electrons, but to a significant extent, with the atomic nucleus as well. Secondly, it continuously picks up electrons that reduce the original charge. A fission fragment, therefore, ionizes more heavily toward the point of its origin, and the amount of ionization it causes actually decreases as the particle losses energy in the absorbing material. Calculation of how far a fission product travels through a particular absorbing material is complicated. However, most fission fragments are stopped by 0.03 m of air at 15°C and standard pressure. Note that this is ten times less than the distance traveled in air by a 5 MeV proton. Hence, fission fragments are completely stopped very close to their point of origin.

1.5.2 Linear Energy Transfer

Both the ICRP and the International Commission on Radiation Units and Measurements (ICRU) have recommended that the effectiveness of a given radiation in damaging tissue be evaluated from its linear energy transfer (LET) for the requirements of radiation protection. The LET of a charged particle in a medium is defined as the average amount of energy locally imparted to the absorbing material by a charged particle of specified energy in traversing a suitably small (ideally infinitesimal) distance within the absorbing material [ICRU, 1970]. It is defined as an average value because the amount of energy imparted by a charged particle to an absorbing material, even over a very small distance may change especially near the end of its path, and also because secondary particles produced by indirectly ionizing radiation, such as photons or neutrons, are not all of the same energy.

LET is a term used extensively when the effect of ionizing radiation on biological material is being considered and its exact meaning and value is thoroughly discussed in the radiation biology literature, see for example, [Pizzarello and Witcofski, 1975]. For our purposes we will simply categorize charged particles as "high" and "low" LET particles in accordance with the general usage of ICRP, ICRU, and NCRP. Since heavy-charged particles lose all their energy in a relatively short path length, they are known as high-LET particles. Light charged particles, like electrons, experience many changes in path direction and lose their energy over a relatively long distance (the ionization produced is dispersed over a longer path or track length). Hence these particles are known as low-LET particles. As for indirectly ionizing particles, photons interacting with matter generally produce light-charged particles and so are low-LET particles, whereas neutrons usually produce heavy charged particles and fission fragments and so are high-LET particles.

1.6 Interaction of Directly Ionizing Particles

Of all the directly ionizing particles, alpha particles and electrons are the most commonly encountered by radiation workers. Radionuclides which produce alpha particles are common in the earth's crust, are frequently used in home smoke detectors, and have been used in the field of radiation oncology. Electrons are commonly produced by the radionuclides used in diagnosis and therapy, are generated by X-ray oncology equipment, and often result from the use of collimation in diagnostic radiology.

1.6.1 Alpha Particles

The only natural source of alpha particles is nuclear decay. Because the alpha particle contains two neutrons and two protons, it has a mass of four units, which makes it a heavy particle. Because it contains two protons, the alpha particle has a positive charge of two times that of the electron, hence the alpha particle is highly interactive. The effect of its large mass and double charge makes an alpha particle highly interactive in the vicinity in which it is produced; hence it never penetrates far into any material. A thin sheet of paper is usually sufficient to stop all but the most energetic alpha particles. When alpha particles travel through a material, they lose energy by collision with atomic electrons and cause ionization to occur. Their large mass and charge result in a path that is, in general, straight and not very long. The maximum distance that is necessary to stop all alpha particles is known as the range. Figure 1.7 illustrates the range of a thin source of alpha particles in tissue of less than 0.04 mm. initially having all the same energy. The mean range is that distance from the source in which the number of alpha particles detected has decreased to one half of the original number detected. As is true of any heavy-charged particle, alpha particles all travel nearly, but not precisely, the same distance in a given medium before coming to rest. This variation in travel distance toward path end is known as straggling and is due to the probabilistic nature of the collisions between alpha particles (and other heavy-charged particles) and atomic electrons. To assess the range of alpha particles, it should be noted that 5 MeV alpha particles are all stopped by 35 mm of air at 15 °C and standard pressure and have a range determined by the distance at which one half of the original number are detected. As is true of any heavy-charged particle, alpha particles all travel nearly, but not precisely, the same distance in a given medium before coming to rest. This variation in travel distance toward path end is known as straggling and is due to the probabilistic nature of the collisions between alpha particles (and other heavy-charged particles) and atomic electrons.

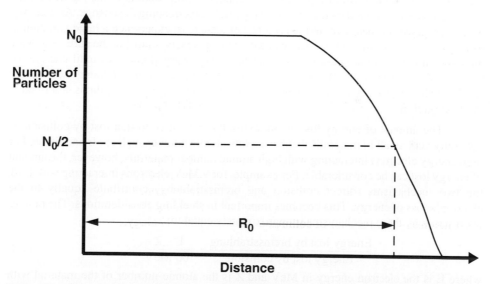

Figure 1.7: Range of alpha particles in matter. N is the number of alpha particles detected at any distance in the material. N_0 is the initial number of alpha particles detected. When N_0 has fallen to half its original value ($N_0/2$) the depth of penetration into the absorbing material at that point is known as the mean range (R_0).

1.6.2 Electrons

The major sources of electrons are 1) nuclear particle (neutron) decay, in which case the resulting electrons are known as (negative) beta particles or negatrons; 2) nuclear de-excitation, in which electrons, known as internal conversion electrons, may be expelled instead of gamma-ray photons; 3) atomic de-excitation, in which electrons, known as Auger electrons, may be expelled instead of characteristic X-ray photons; and 4) the end product in a material's attenuation of photons having energies in the visible-light region and above (see discussion of photon interactions with matter below). In the case of beta decay and most photon interactions (the photoelectric interaction case is a notable exception) the electrons do not have a specific energy when they are produced so, in general, one is dealing simultaneously with a number of electron energies limited by some maximum energy value. Electrons are very light particles, and have a negative charge.

Because of its light mass and single charge, an electron is not as interactive as an alpha particle and, therefore, has a much longer range in matter. Also, electrons tend to travel through matter in tortuous paths rather than in straight lines. In addition to losing energy by collision with atomic electrons, another mechanism by which electrons can lose energy (in the presence of a nucleus) is a braking action known as bremsstrahlung. Whenever a charged particle undergoes a change in direction or magnitude of its motion, it emits energy in the form of a photon. This change is proportional to the nuclear charge Z, which causes it, divided by the mass of the particle experiencing the change. An electron, because of its small mass can, in the presence of a nucleus, experience a large variation in its motion. Since the amount of change of motion is inversely related to the particle's mass, bremsstrahlung does not become important for heavy-charged particles until energies of billions of electron volts (GeV) are reached. Whenever bremsstrahlung occurs, the resulting photon produces charged particles (mostly electrons) by the methods described in the next section.

The amount of energy lost by bremsstrahlung, relative to that lost by collision, is generally very small in materials that have a low atomic number such as air and tissue. For high energy electrons interacting with high atomic number materials, however, the amount of energy lost can be considerable. For example, for 9 MeV electrons interacting with lead, the two mechanisms (direct collision and bremsstrahlung) contribute equally to the electron's loss of energy. This becomes important in shielding considerations. The ratio of energy loss by these mechanisms is approximately equal to:

$$\frac{\text{Energy lost by bremsstrahlung}}{\text{Energy lost by collision}} \sim \frac{E \cdot Z}{700 \text{ MeV}} \tag{1.3}$$

where E is the electron energy in MeV and Z is the atomic number of the material with which it is interacting.

The range of electrons in matter is a function of both the maximum energy of the electrons and the density of the material through which the electrons are traveling. Density, symbolized by the greek letter ρ, is defined as mass per unit volume of material. It is possible to obtain a relationship between the range R of electrons in a particular material and the density ρ of that material. In fact the range of an electron in a material times the density is equal to a constant number k. In mathematical form:

$$R \cdot \rho = k \qquad (1.4)$$

The quality k is often termed the density thickness. Certain physical properties including density thickness, of some commonly produced electrons are given in Table 1.2 Use of Eq. (1.4) is illustrated in the following example.

Example 1.1

Calculate the minimum thickness of aluminum necessary to stop all the electrons from a phosphorus-32 (^{32}P) source. From Table 1.2, the maximum electron (beta particle) energy is 1.71 MeV and the density thickness of these electrons in unit density material is 8.0 kg· m^{-2}. The density of aluminum is 2.7·10^3 kg·m^{-3}. Using Eq. (1.4):

$$R \cdot 2.7 \cdot 10^3 \text{ kg} \cdot \text{m}^{-3} = 8.0 \text{ kg} \cdot \text{m}^{-2}$$

Solving for R:

$$R = \frac{8.0 \text{ kg} \cdot \text{m}^{-2}}{2.7 \cdot 10^3 \text{ kg} \cdot \text{m}^{-3}} = 2.96 \cdot 10^{-3} \text{ m} \sim 3 \cdot 10^{-3} \text{ m} = 3 \text{ mm}$$

From the above example we see that in this case a few millimeters of aluminum is sufficient to stop even the most energetic electrons produced by such a radioactive source.

Table 1.2: Selected Physical Properties of Commonly used Electron Sources.

Source	^{3}H	^{14}C	^{35}S	^{32}P
Maximum electron energy (MeV)	0.018	0.154	0.167	1.71
Average electron energy (MeV)	0.006	0.05	0.049	0.70
Density thickness (k) in unit-density material (kg · m^{-2})	0.0052	0.29	0.32	8.0
Fraction transmitted through dead layer of skin (0.07 mm)	0.0	0.11	0.16	0.95

1.7 Interaction of Indirectly Ionizing Particles

Of all the indirectly ionizing, i.e., uncharged, particles, photons and neutrons are the most commonly encountered by radiation workers. Neutrons are included here because they are finding application in radiation oncology and are also of great concern to many industrial based physicists, engineers, and technologists.

1.7.1 Photons

The concept of electromagnetic radiation embraces a large class of physical phenomena: radio and television waves, microwaves, visible light, ultraviolet, X- and gamma ray photons. Although ultraviolet photons are sometimes encountered by radiation workers, X- and gamma-ray photons are the only types of photons that are of interest in the context of this book. Consequently, the photon energy range considered is between 10 and 50000 keV (0.01 to 50 MeV). It is important to keep in mind that the difference between X- and gamma-ray photons is in origin, not energy. The term X-ray photons is applied to those photons produced by interactions involving inner-shell atomic electrons or by bremsstrahlung. On the other hand, the term gamma-ray photon is applied when photons are the result of interactions involving the atomic nucleus. Because of the amount of energy associated with them, X- and gamma-ray photons interact with matter in special ways. An understanding of each of these interactions is helpful in realizing why X- and gamma-ray photons are potentially harmful, how they are detected in radiation protection work, and why the particular methods used to protect against them are effective. In this book we are concerned only with the three most common interactions of photons with matter.

The first method by which X- and gamma-ray photons interact with matter is known as the photoelectric effect (Figure 1.8). In this type of interaction a photon has a particle–particle collision with an atomic electron, and the photon transfers all of its energy to the electron. If that energy is sufficient to release the electron from its atomic orbit, the atom is ionized. The energy of the electron is equal to the total energy of the photon minus the binding energy of the released electron. The freed electron interacts in the manner described in Section 1.6.2.

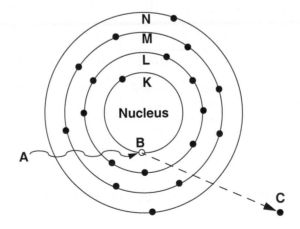

Figure 1.8: The Photoelectric Effect: The incoming photon A has energy equal to or greater than an electron B in the K shell. All the energy of photon is transferred to the electron releasing the electron C from the atom. The released electron has energy equal to that of the photon minus the binding energy of the K shell.

A second method by which X- and gamma-ray photons interact with matter is known as the Compton effect (Figure 1.9). In this type of interaction a photon collides with an atomic electron. This time, however, it transfers to the electron only part of its energy. The rest of the original photon's energy is radiated as a lower-energy photon. Usually, this secondary photon travels in a different direction from the one creating it. This action is referred to as scattering. It is possible for the lower-energy Compton photon to undergo either a photoelectric or a further Compton interaction. In the case of a Compton interaction, the electron is viewed as "free", i.e., the binding energy of the electron in its atomic orbit is insignificant when considered relative to that of the interacting photon. The energy of the original photon is divided between the electron and the secondary photon according to a complex but fixed set of rules. The electron released by the original Compton interaction, and any further electrons released, interact as previously described.

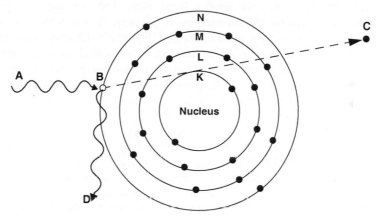

Figure 1.9: The Compton Effect: The incoming photon A has an energy much greater than the binding energy of the electron B. When the photon interacts with the electron B, it transfers only some of its energy to the electron. The energy transferred to the electron C releases it from the atom. The remainder of the photon's energy is re-radiated as a photon D of lower energy.

The third method by which X- and gamma-ray photons interact with matter is known as pair production (Figure 1.10). In this type of interaction the photon, in the presence of a nucleus, disappears, changing all its energy into matter in the form of an electron and a positive electron, known as a positron. In order for this interaction to occur, the photon must have an energy of at least 1.02 MeV, i.e, an energy at least equal to the rest mass (see Glossary) of the two particles produced. The total energy of the photon is shared equally between the two particles, giving them mass and (perhaps) energy. The electron then interacts as before. The positron loses its extra energy, if any, by ionization. When the positron has lost all its energy, it unites with an electron and the two particles disappear, or annihilate. Two characteristic annihilation photons are produced (see Figure 1.10). These annihilation photons move away from each in opposite directions at almost exactly 180°, and have the same energy (0.511 MeV). These photons then interact with matter by either the photoelectric or Compton effect.

Although photons do undergo other types of interactions with matter, at the photon energies encountered by most radiation workers, the three interactions just discussed are the most important ones. Which type of interaction a photon will undergo is determined by two factors: the energy of the photon and the material it is traversing. In general, photons of lower energy (less than 100 keV) are more likely to interact through the photoelectric effect; photons of intermediate energy (above 100 keV) through the Compton effect. Pair production is not possible until photon energies of 1.02 MeV or greater are reached. Photons with energy at or above 7 MeV, can interact directly with the atomic nucleus by a process known as photodisintegration which frequently results in the production of neutrons. This often happens when very high energy particle accelerators are used for medical (usually radiation oncology) or other purposes (such as high energy physics research). The specific energy range of each interaction is also determined by the properties of the material through which the photon is traveling. For example, pair production becomes dominant in lead for photon energies in the range of 4 to 5 MeV, whereas, for tissue, photon energies of 20 MeV must be reached. Figure 8.5 (Chapter 8) illustrates how the probability (known technically as the interaction cross-section) of the three major photon interactions changes in lead as a function of photon energy. Whenever a beam of photons traverses a material, some of the photons interact with the material and are lost from the beam. This process is referred to as *attenuation*. To attenuate a photon beam simply means removing some of the photons from it by interactions between the photons and the material. When a photoelectric interaction occurs, all of the photon's energy is absorbed or attenuated out of the beam. When a Compton interaction occurs, some of the photon's energy is absorbed out of the beam and the rest is scattered from the beam in the form of the secondary photon produced. In pair production, the photon's energy is absorbed or attenuated out of the beam. Photons can also be attenuated out of the beam by coherent or Rayleigh scattering, in which case no energy is transferred to the material, the photon simply changes its direction of motion.

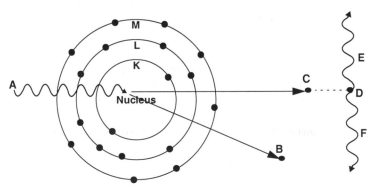

Figure 1.10: Pair Production: A photon A of energy equal to or greater than 1.02 MeV disintegrates, in the presence of a nucleus, into an electron B and a positron C. The positron loses energy by ionization until it finds an electron D and annihilates, producing two characteristic annihilation photons E and F. Note that the annihilation photons travel in directions almost exactly opposite to each other.

1.7.2 Neutrons

Neutrons have a mass comparable to that of a proton, but they are uncharged (i.e., electrically neutral). Hence the behavior of neutrons in matter is quite different from that of either charged particles or photons.

Neutrons are primarily produced in photodisintegration, as an end product in a nuclear interaction, or in a nuclear fission event. As an example of a nuclear interaction consider that of polonium (Po) with beryllium (Be). Polonium emits an alpha particle that enters the nucleus of the beryllium atom and remains there for a very short time, forming what is known as a compound nucleus. After a time interval of about 10^{-20} to 10^{-12} s, a neutron having an energy of about 14 MeV is emitted. Symbolically this reaction looks like:

$$_4^9\text{Be} + {}_2^4\text{He} \Rightarrow \text{Compound Nucleus} \Rightarrow {}_6^{12}\text{C} + \text{n}$$

$$_4^9\text{Be}(\alpha, \ \text{n}) \ {}_6^{12}\text{C} \tag{1.5}$$

The second form of Eq. (1.5) uses the notation (α,n), generically written as (A,B). In this notation A denotes the incoming particle, an alpha particle symbolized by α in the case of Eq. (1.5), and B denotes the outgoing particle, a neutron symbolized by n in the present case. Figure 1.11 illustrates a nuclear interaction. Notice that beryllium is a very low atomic number element.

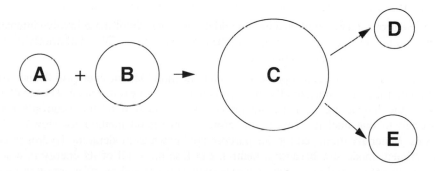

Figure 1.11: A neutron A interacts with a nucleus B, and forms a compound nucleus C. A short time later, 10^{-12} s or less, particle D is expelled and a residual nucleus E is left.

Based on the amount of energy they have, neutrons are generally divided into two categories known as fast and thermal. Since fast is a very large energy range (i.e., everything above 0.025 eV at room temperature), neutrons with energies below 0.5 eV are sometimes called slow neutrons (see Glossary). The energy 0.5 eV is used because neutrons with energies below this are readily absorbed by cadmium and neutrons with energies above this value are not.

A fission event occurs when a very heavy nucleus breaks into two large nuclear fragments. Fission can occur spontaneously, although such events are rare, or it can be triggered by a (usually) very low-energy (known as thermal) neutron (about 0.025 eV at room temperature) entering the nucleus of a very high atomic number element like $^{235}_{92}U$ and forming a compound nucleus, which after a short time, breaks into two large nuclear fragments. Whenever fission occurs, one or more neutrons are produced and usually a large amount of energy as well. These neutrons generally have a high energy and are known as fast neutrons. The exact nuclear fragments into which the nucleus disintegrates is largely a probabilistic outcome. An example is:

$$^{235}_{92}U + n \Rightarrow \text{Compound Nucleus} \Rightarrow {}^{96}_{36}Kr + {}^{137}_{56}Ba + 3n \qquad (1.6)$$

Notice that in Eq. (1.6), three neutrons are produced and that the two fission fragments produced are quite disparate in mass number. In any fission event there is always a "light" fission fragment ($^{96}_{92}Kr$) and a "heavy" fission fragment ($^{137}_{56}Ba$) produced. Fission is most probable for ^{235}U, ^{239}Pu, and ^{233}U for thermal neutrons. However, some nuclei such as ^{238}U will absorb fast neutrons and fission, but the probability of this happening is very small. Usually when ^{238}U absorbs a neutron, it becomes ^{239}U, which decays via beta minus decay to ^{239}Np, which in turn decays by release of another beta minus decay to ^{239}Pu.

Neutrons interact with matter in a variety of ways. All of these involve interaction with the atomic nucleus, however, and not the atomic electrons. We shall briefly describe each of the most commonly encountered.

Elastic scattering is that process in which the initial kinetic energy of the neutron is shared with the atomic nucleus. The nucleus is not left in an excited state. The smaller the mass of the interacting nucleus, the greater is the fraction of the neutron's energy transferred to that nucleus. This is, therefore, a very good method for slowing down neutrons. Materials employed for this purpose are known as moderators. Hydrogen is the most efficient moderator because a neutron can lose up to all of its energy in a single collision; therefore, hydrogen-rich materials such as water and paraffin are often used as moderators. Other good moderating materials are heavy water (water in which deuterium atoms, ^2_1H have been substituted for some of the hydrogen atoms, ^1_1H), beryllium, and carbon.

In the process of inelastic scattering, possible only for fast neutrons, the atomic nucleus is left in one of its excited states. In most instances, the nucleus quickly radiates this excitation energy in the form of a gamma-ray photon. This photon production is an important consideration when designing neutron shielding, but it is an unwanted complication in detecting neutrons. When the incident neutron energy is 10 MeV or higher, it is possible for a second neutron to be radiated by the nucleus. Thus in inelastic scattering we generally have (n,nγ) which signifies a neutron before the interaction and a neutron plus a photon afterwards. Or, the interaction may be described as (n,n) or (n,2n) with no photon produced and the nucleus remaining in a metastable excited state.

The most common interaction of neutrons with matter is simple capture. This consists of neutrons being captured by a nucleus and the subsequent production of a gamma-ray photon that usually has energies of several MeV. Thermal neutrons induce this reaction (n,γ) in nearly all nuclei. There is also a high probability for this to occur in some nuclei at particular neutron energies above the thermal range; this is known as "resonance" capture. Cadmium has a very high probability of interacting by simple capture (n,γ) with neutrons having energies up to 0.5 eV. Cadmium, therefore, is useful for detecting and classifying neutron energies and is used quite extensively as a material in control rods in a nuclear reactor or pile, (a machine in which controlled nuclear fission take place) but, because of the production of photons, cadmium is not so useful as a shielding material.

Often when neutrons are absorbed by atomic nuclei, charged particles are ejected instead of, or in addition to, gamma-ray photons. This interaction is most probable in the case of a light atomic nucleus and a fast neutron. There are important exceptions to this rule; in certain cases alpha particles are produced by thermal neutrons. Examples are:

$$ {}^{6}_{3}\text{Li} + \text{n} \Rightarrow {}^{7}_{3}\text{Li} \Rightarrow {}^{3}_{1}\text{H} + \alpha $$

$$ {}^{6}\text{Li}(\,\text{n}, \alpha)\,{}^{3}\text{H} $$

$$ {}^{10}_{5}\text{B} + \text{n} = {}^{11}_{5}\text{B} = {}^{7}_{3}\text{Li} + \alpha $$

$$ {}^{10}\text{B}(\,\text{n}, \alpha)\,{}^{7}\text{Li} \tag{1.7} $$

The boron interaction above, with the subsequent detection of the produced alpha particle, is a favored method for identifying thermal neutrons. In addition, because of the production of an alpha particle, boron is used in neutron therapy for certain types of brain tumors. This type of therapy has had mixed success and is also hampered by the need to have a source of thermal neutrons present. This usually implies that the neutron therapy facility must have nuclear (fission) reactor on the premises.

The fission interaction takes place only with heavy atomic nuclei. The neutron is absorbed and the resulting compound nucleus splits into two fission fragments and one or more neutrons as previously described.

Finally, the capture by a nucleus of neutrons having energies of 100 MeV or higher, may cause the emission, known as a shower, of many different types of particles.

1.8 Review Questions

1. What accounts for more than 90% of the total man-made radiation exposure to the world population?

2. Is radiation from nuclear power plants distinguishable from background radiation?

3. What percentage of the total radiation exposure is attributable to occupational radiation exposure?

4. Is there a need for training for persons would regularly work with sources of ionizing radiation? State the reason(s) for your answer.

5. What are the two major categories of natural background radiation?

6. What are the constituents of the nuclear atom.

7. What are the three basic forms of radioactive decay? How do the three forms of beta decay differ?

8. Describe the process of ionization.

9. What are the major types of ionizing radiation?

10. To what process is internal conversion is an alternative process?

11. What is the difference between a 600 keV X ray photon and 600 keV gamma ray photon?

12. Assuming that the transition energy is high enough, to what process is electron capture an alternative?

13. For which radioactive decay mode does the mass number (A) change?

14. Of what three categories or types of particles does directly ionizing radiation consist? Describe the manner in which each category interacts with matter.

15. What are the major indirectly ionizing particles?

16. Define the term linear energy transfer (LET).

17. What is the meaning of the term "mean range" for alpha particles?

18. What is the meaning of the term "range" applied to electrons?

19. What is "bremsstrahlung"? Briefly describe this mechanism for an electron.

20. What are the three most common ways in which photons interact with matter?

21. Into what two categories (based on energy) are neutrons divided?

22. Neutrons usually interact with matter by forming a compound nucleus. Describe what this is.

23. What is the meaning of the term radioactivity? What is the meaning of the term half-life?

24. What is the SI unit of radioactivity? How is it defined?

1.9 Problems

1. Consider an atom in which the binding energy of the "K" shell electrons is 30 keV and the binding energy of the "M" shell electrons is 0.7 keV. An electron with a kinetic energy of 25.3 keV is found to have been ejected from the "M" shell of this atom. The ejection was caused by Auger electron emission following an "L" to "K" transition. What is the binding energy of the "L" shell electrons?

2. ^{224}Ra spontaneously decays via one of two alpha particles which have energies of $\alpha_1 = 5.7837$ and $\alpha_2 = 5.5427$ respectively. The resultant nuclide ^{220}Rn is produced directly in its ground state when α_1 occurs. There is direct isomeric transition of ^{220}Rn to its ground state when ^{224}Ra decays into α_2. What is the energy of this isomeric state? What is energy of the gamma-ray photon that is released?

3. ^{22}Na spontaneously decays alternatively via electron capture and two possible positron emissions into ^{22}Ne. One positron has a maximum energy of 1.8210 MeV and other of 0.5460 Mev. The higher energy positron results in ^{22}Ne being produced in the ground state of the nucleus. Emission of the lower energy positron results in ^{22}Ne being produced in the one excited state of the nucleus. What is the difference in energy between the two nuclear states of ^{22}Ne? What is the energy of the isomeric (gamma-ray photon) transition? What must be the minimum energy mass defect difference between ^{22}Na and ^{22}Ne for positron emission to occur? When the positron in not emitted with the maximum energy possible, what happens to the remaining energy?

4. What is the ratio of energy lost through radiation in the form of bremsstrahlung to energy lost by collision when an electron traveling through aluminum has decreased its energy to 7 MeV? What if the electron were in the lead?

5. What is the range of beta particles emitted by ^{14}C traveling through aluminum?

6. What is the range of beta particles emitted by ^{32}P traveling through lead?

7. An 80 keV photon undergoes a photoelectric interaction releasing a K shell electron from a tungsten absorber. The binding energy of an electron in the K shell of tungsten is almost 70 keV. What will be the kinetic energy of the released electron? Suppose the energy of the incoming photon had been 60 keV. Would photoelectric interaction with the K shell electron have been possible? Suppose

three or four photons, each of energy 60 keV were to bombard the K shell electron of tungsten, would an electron be released?

8. A 200 keV photon undergoes a Compton interaction in tissue. If the secondary photon has an energy equivalent of 150 keV, how much energy was transferred to the electron?

9. A 2 MeV photon undergoes pair production, dividing its excess energy equally between the produced electron and positron. What is the kinetic energy of the electron?

10. In a fission reaction, $^{235}_{92}$U decays into $^{131}_{53}$I. What is the other fission product which is produced? How many neutrons are released?

Chapter 2

DETECTORS

Every radiation detection system consists of two parts. The first is the detector proper, in which the interaction of radiation with matter takes place. The second is the measuring apparatus, sometimes called the counter or readout component. This component takes whatever the detector produces and performs those functions required to accomplish the desired measurements.

Different types of systems are characterized by the nature of the interaction of radiation with the detector. Several types operate by virtue of the ionization produced in the detector by the passage of charged particles. Examples of such systems are ionization chambers, proportional counters, Geiger counters, semiconductor radiation detectors, and cloud and spark chambers. Others, such as scintillation detectors and thermoluminescent devices, operate by virtue of exciting atomic electrons to higher energy states and the readout is based on the prompt or delayed emission of photons as the electrons return to their normal energy state.

If the primary radiation consists of charged particles, the interaction with the detector is produced directly. If the primary radiation consists of uncharged particles, i.e., neutrons or photons, the interaction originates by secondary processes. Both the direct and indirect interaction process are described in Chapter 1.

Detection systems, as a whole, are classified by their mode of operation, i.e., whether they are pulse-type or mean-level devices. In the pulse-type mode of operation, the output of the detector is a series of signals, or pulses, resolved or separated in time. Each signal represents the interaction of a nuclear particle with the detector. A Geiger counter is an example of a detection system that usually operates in pulse mode.

In the mean-level or integrating mode of operation, the quantity measured directly is the average effect due to many interactions of radiation with the detector. No attempt is made to resolve individual events. Frequently, such resolution is impossible because of the high rate at which the interactions occur. An ionization chamber is an example of a detection system that is most often operated in mean-level mode.

2.1 Gas-Filled Detectors

Those systems in which detectors operate by virtue of charged particles producing ionization in a gas-filled chamber are known as gas-filled detectors. The three most common systems of this type are ionization chambers, proportional counters, and Geiger counters.

Detectors associated with each of these systems employ a gas-filled chamber with a generally (there are some exceptions) positively charged central electrode in the form of a thin wire insulated from the negatively charged electrode which is usually formed by the chamber walls. The positively charged electrode is known as the anode, and the negatively charged electrode as the cathode. A voltage (V) applied between the wall and central electrode, maintains the charge distribution. A typical gas-filled detector is shown schematically in Figure 2.1. As a charged particle passes through the gas it ionizes atoms and forms within the chamber, ion pairs composed of the residual positively charged atoms and the negatively charged electrons. The positive and negative charges thus produced have a force applied to them by the voltage; positively charged particles tend to move toward the chamber walls and negatively charged particles toward the central electrode, because, of course, opposite charges attract. The central electrode is often referred to as the collector.

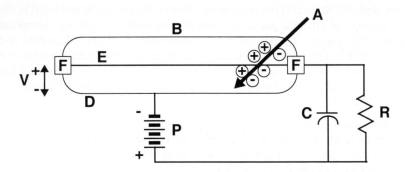

Figure 2.1: A Gas-Filled Detector: The incoming ionizing radiation A enters the detector B and causes ionization. The residual positive ions move toward the chamber walls D, which are negatively charged, and thus form the cathode. The electrons (negative ions) move toward the central wire E, which is positively charged and thus forms the anode or collector. The central wire and the chamber walls are electrically isolated from each other by the insulating material F. The voltage level V is maintained by the power source P. The resistor R and capacitor C circuit integrates individual pulses from the electrons.

The main differences among the three types of gas-filled detectors lie in the gas used, the pressure at which that gas is maintained within the chamber, and the voltage level that is maintained between the central electrode and walls of the chamber. Figure 2.2 is a graph of the voltage versus the amount of ionization that occurs.

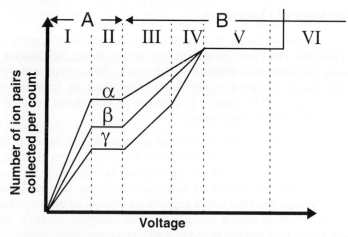

Figure 2.2: Voltage Dependence of a Gas-filled Detector: Region I represents the voltage levels at which partial recombination of ion pairs take place. Region II is the ionization chamber region. No recombination of ion pairs takes place here, and no secondary electrons are produced. Region III is the region of proportionality. The number of secondary ion pairs produced is directly proportional to the number of primary ion pairs produced. Region IV is the region of limited proportionality. Region V is the Geiger region. Here the maximum number of ion pairs is produced by even one primary ion pair. Region VI is the region of continuous discharge. **A** represents voltage levels that produce simple ionization. For the voltage levels **B**, multiplication of primary ionization takes place.

2.1.1 Ionization Chambers

Ionization chambers are used as survey meters, as dose calibrators, as personnel monitors, and as standard measuring devices for determining radiation levels. They are useful because they can indicate the quantity of the incoming radiation and in some cases its energy as well. This capability enables the user to identify the source as well as the strength of radiation present as in the case of a radionuclide, or to calibrate the beam quality and quantity of an X-ray generator.

The detector proper can take many forms. The most common is that of a cylindrical conducting chamber containing a central electrode located on the axis of the chamber and insulated from it (see Figure 2.1). The usual gas filling for an ionization-chamber detector is dry air at atmospheric pressure (101.3 kPa). Other gases are sometimes used for particular purposes, but in radiation protection work, air is commonly used.

An ionization chamber can be used to some extent for detection of all types of particles. However, in radiation protection work it is most commonly used for detection of both photons and high-speed electrons. In the mean-level mode of operation, the ionization chamber can be used in two ways: (1) as a current-measuring device yielding a quantity proportional to the rate (amount per unit time) of arrival of the ionizing radiation within the detector, and (2) as a voltage-measuring device yielding a quantity related to total radiation incident on the chamber during the entire period of measurement.

2.1.1.1 Principles of Operation

An ionization chamber demands that only ionization produced directly by incoming radiation be collected. Therefore, for quantitative measurements, the detector must furnish an output having a definite relationship to this ionization. To accomplish this goal, a known fraction of the charge produced within the chamber must be collected and no other ionization can take place. Therefore, it is necessary to understand how the electron and residual positive ions, once formed, behave.

When an outer-shell or valence electron is released from an atom, the immediate inclination of electron and positive ion is to reunite. To avoid this "recombination" a certain minimum level of voltage must exert enough force on the electron to separate it from the positive ion and to send both of them on their respective ways. Generally the walls of the chamber are the cathode; i.e., they have a negative potential or an excess of electrons. Thus the positive ions, traveling slowly because they are relatively massive, migrate to the chamber walls, where they receive an electron and are neutral once again.

The central wire is usually the anode, which is positive with respect to the chamber walls; hence electrons are attracted to it. Electrons, however, are very light, and a small force applied to them (by the voltage V in the detector) gives them a lot of energy. Therefore, the voltage applied in the chamber must be low enough to keep the electrons from acquiring so much energy before reaching the anode that the electrons themselves ionize other atoms. This type of ionization is called secondary ionization. Usually ionization chambers use voltages between 100 and 300 V, although some have higher operating voltages.

2.1.1.2 Electron Movement to Collector

It is illuminating to describe how electrons, once formed, move through the surrounding gas atoms to the collector electrode (see Figure 2.3). When an electron travels toward the anode, it does so by a crooked path because it collides with the neutral atoms located everywhere in the chamber. The more atoms with which an electron can collide, the more it will do so, and the shorter will be the distance that the electron travels between collisions; also the electron will acquire less energy between collisions. Keep in mind that the voltage applies a force to the electrons, thus giving them energy.

The average distance that an electron travels between collisions is called the mean free path, which is inversely proportional to the pressure of the gas, i.e., to the number of gas molecules per unit volume. When an electron collides with a gas molecule, it is deflected from its original path. The direction of the electron's motion is random; however, the net drift of the negative electron is toward the positive anode. The speed with which the electron drifts toward the anode is known as the drift velocity; it is directly proportional to the applied voltage in the chamber and inversely proportional to the pressure.

One more property of electrons must be taken into account. Electrons have a tendency to attach themselves to neutral atoms, thus forming negative ions. Air, for

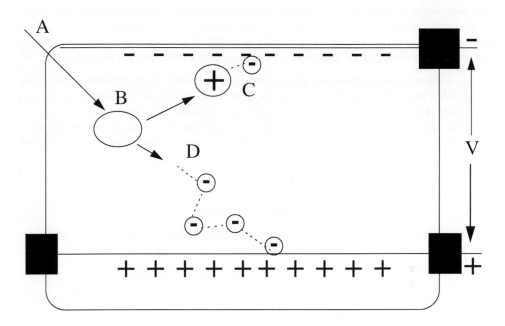

Figure 2.3: When ions are produced in the gas (usually air) which fills the ionization chamber, the electrons are attracted by the anode represented here by a thin central wire. The heavier, positive ions travel toward the cathode, represented here by the chamber walls, attract an electron, and become neutral gas atoms once again. Incoming ionization radiation A, interacts with air molecule B to produce a positive ion C and a negative electron D. The electron then follows an irregular path to the anode.

example, has a relatively high probability of attracting free electrons. Since the negative ions thus formed are massive and slow and do not drift toward the anode at the same rate as the electrons, the characteristics of the gas-filled detector are changed. This phenomenon is not a serious problem with respect to ion chambers, but it is with Geiger counters. Hence, this property of electrons must be neutralized by using a gas whose atoms have a low electron affinity, i.e., do not attach electrons to themselves very readily. The inert gases such as helium, argon, neon, and krypton have a low electron affinity and, therefore are, as discussed below, preferable as filling gases for Geiger counters.

The use of ionization chambers as survey and calibration meters and personnel monitors is discussed in Chapters 5 and 6.

2.1.2 Geiger Counters

Geiger counters, also called Geiger-Müller or G-M tubes, are widely used as monitoring instruments because they can detect any type of radiation that produces ionization within the detector, no matter how small the amount of ionization. In fact, this high sensitivity makes Geiger counters ideally suited for detecting fast electrons and photons because the low probability of these particles producing ionization makes them hard to detect. Alpha particles and other highly ionizing particles can be detected by G-M tubes, but because of their short range they are unable to penetrate the walls of the detector unless thin windows are used or the source of the radiation is placed inside the detector (see Section 2.1.4).

Geiger counters are almost always used in pulse mode. Their operation requires that the number of electrons collected at the anode be independent of the amount of primary ionization. Hence this factor cannot be used as a measure of particle energy, nor is it possible to discriminate among different types of particles by means of the sensitivity of the electronic circuit. Therefore, G-M tubes can tell nothing about the energy of radiation.

The great utility of the G-M tube is the result of several factors:

1.   It has very high sensitivity.

2.   It can be used with different types of radiation.

3.   It can be fabricated in a wide variety of shapes; i.e., the width can be anywhere from 2 mm to several centimeters, and the length, 1 cm to several meters.

4.   Its output signal is so strong that it requires little or no amplification.

5.   It is reasonable in cost, because the tube, as well as the electronics following it, are extremely simple in construction.

2.1.2.1 Principles of Operation

The tube itself is generally a metallic envelope in which two electrodes and a proper filling gas are incorporated. The internal, or collector, electrode (anode) is a fine wire (about 0.10 mm in diameter) frequently made of tungsten because of its strength and uniformity in small-diameter applications. The anode is most often supported at both ends with insulators (see Figure 2.1). Sometimes, however, the anode is terminated at one end by a glass bead or a loop. As opposed to ionization chambers, the most common filling gas for a Geiger counter is one of the noble gases: helium, argon, or neon. The only requirement is that the gas have a small electron affinity. A wide range of pressures is used, the most frequent being 9 to 50 kPa.

In order to achieve the high sensitivity of the G-M tube, a relatively high operating voltage must be used (generally between 900 and 1200 V). Again this differs from that of the ionization chamber.

The quality of a G-M tube is determined by the counting rate versus voltage response, as shown in Figure 2.4. That range of operating voltages through which the counting rate changes very little is called the plateau. A good tube has a counting-rate change of less than 10% throughout the plateau region. The voltage at which the G-M tube begins to count is known as the starting voltage. Voltages higher than the plateau voltage result in a breakdown of the tube gas; i.e., ionization occurs because the voltage itself pulls the gas atoms apart even without the presence of any ionizing radiation.

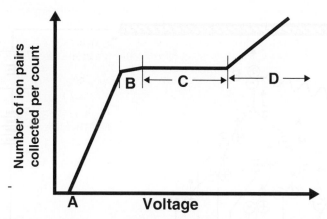

Figure 2.4: Geiger Counter Plateau: **A** represents the starting potential or starting voltage, **B** represents the "knee" or threshold voltage, **C** represents the plateau region, and **D** represents the breakdown or continuous discharge region.

2.1.2.2 Count Method

The production of counts in a G-M tube occurs in the following manner: Electrons produced by the primary ionizing event drift toward the anode, but lose energy when they are scattered away from it. The electrons do not gain enough energy between collisions to produce secondary ionization until they arrive within the last few mean free paths of the anode. Once ionization starts, it builds up rapidly, since secondary ionization events can produce more ionization. This buildup is known as a Townsend avalanche (Figure 2.5).

The initial Townsend avalanche terminates when all the electrons associated with it have reached the anode. However, the initial avalanche is followed by a succession of avalanches, each of which is triggered by the preceding one. The termination of the avalanches comes when the number of positive ions surrounding the anode reduces the voltage of the anode to a level that is inadequate for maintaining ionization. The entire production of avalanches occurs in a fraction of a microsecond. In this short time electrons are collected at the anode; then the positive ions migrate to the cathode.

It must be kept in mind that the voltage across the G-M tube is high; hence the positive ions are able to gain a lot of energy on their way to the cathode. The collision of these energetic positive ions with the cathode causes either of two effects:

1. An electron is ejected from the cathode with an amount of energy that is the difference between the energy of the positive ion and the ionization potential (work function) of the atoms composing the walls of the cathode. When the electron binds to the positive ion, this energy difference plus the binding energy is radiated as a low energy photon, which subsequently produces a photoelectron.

2. More than one electron can be ejected directly from the cathode by a single positive ion.

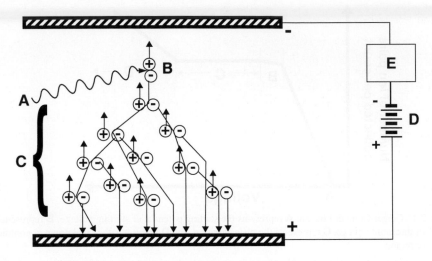

Figure 2.5: Production of Avalanches in a Geiger-Müller Tube: The incoming ionizing radiation **A** forms a primary ion pair **B**. The electron thus released produces many secondary ion pairs **C**. A very high potential difference (voltage **D**) is applied to the tube, and the ionizing events are recorded by the measuring apparatus **E**.

In either case, these "extra" electrons start another series of avalanches if they are not stopped. The procedure used to prevent positive ions from producing further avalanching is called "quenching".

2.1.2.3 Quenching

Three methods exist for quenching a G-M tube. The first is known as electronic quenching. In this situation, once avalanching at the anode has stopped, the voltage across the tube itself is lowered. Thus the positive ions do not gain a sufficient amount of kinetic energy to produce excess electrons or photoelectrons. However, this very old method of quenching has the disadvantage that, because of the lowered voltage, positive ions take longer to get to the cathode, so a longer time elapses before the tube is ready to accept another count. Also the electronics connected with the tube must be more complicated, thus raising the cost of the Geiger counter as a whole.

The second method is known as organic quenching. Here, a small amount (about 10%) of organic gas is mixed with ordinary tube gas. Although any alcohol gas will do, a common organic gas used is ethanol (CH_3CH_2OH, molecular weight 46.07). The positive ions, on their way to the cathode, bump into these large, electron-rich organic molecules and transfer their charge to them. The organic molecules, being massive and hence slow, migrate to the cathode and release electrons. The energy, however, goes to dissociating the organic molecules rather than to starting new avalanches. When all the quenching gas has been dissociated, the G-M tube is no longer quenched and hence no longer useful. Approximately, 10^9 organic molecules are dissociated in each set of avalanches, so an organic quenched G-M has a finite lifetime of about 10^{10} counts. Because of its finite lifetime, this method of quenching is infrequently used.

The third method is known as halogen quenching. With this procedure, about 0.1% halogen gas, usually chlorine or bromine, is added to the tube gas. The halogens readily give up an electron to neutralize the positive ions. When the halogen gas gets to the cathode, the extra energy released goes into dissociating the halogen molecule. However, halogen atoms quickly reassociate. Thus the lifetime of a G-M tube that is halogen-quenched is practically infinite. Also it is possible to run the tube at a higher voltage. One precaution must be observed with halogen quenching: The material composing the walls of the tube must not absorb the halogen used.

The sensitive volume of the tube is that volume in which, when a primary ionizing event occurs, an avalanche will result. Halogen-quenched tubes have a larger sensitive volume than, for example, organic quenched tubes.

2.1.2.4 Resolving Time

Since the G-M tube is a pulse-type device, its resolving time, i.e., the time between counts, must be discussed. The only restriction on the counting system is that it register an avalanching sequence when it occurs in the tube. Once a primary ionizing event starts a series of avalanches, a new primary ionizing event is not detected until the avalanching stops and the positive ions are neutralized.

In general the dead time is defined as the time from entry of an initial ionizing event until avalanching at the anode terminates; the recovery time, as the time required for complete recovery of the electronics after the end of the dead-time interval; and the resolving time, as the sum of dead time plus recovery time. A good Geiger counter has an electronic circuit that gives a resolving time very close to the dead time. The dead time varies from count to count even in the same G-M tube, but it is typically 100 to 200 µs. The use of Geiger counters as survey meters and personnel monitors is discussed in Chapters 5 and 6.

2.1.3 Proportional Counters

In the previous section we discussed the production of secondary and continued ionization near the anode. In that previous case, virtually all the gas atoms in the sensitive volume of the detector were ionized each time even one ion pair was created within it. All dependence on the energy of the ionizing particle was lost. Referring back to Figure 2.2, we see there is a voltage region between the ionization chamber region and the G-M tube region designated as the proportional region. A very useful detector, known as a proportional counter can be designed to operate in this region. Multiplication of ionization still takes place, but a strong dependence on particle energy is maintained. These detectors can be operated either in pulse or mean-level mode, but their major application is in pulse mode. The proportional counter offers an important addition to the gas-filled detectors previously discussed, because their output signal is much higher than that produced by the ionization chamber, but the ability to distinguish energy characteristics is maintained.

2.1.3.1 Principles of Operation

The production of counts in the proportional counter differs from that in the G-M tube in two important aspects. In the G-M tube, ionization is terminated when the number of positive ions near the anode is so great that it reduces the voltage of the anode to a level inadequate for maintaining ionization. In the proportional counter, ionization stops when all of the electrons within the local gas volume are swept to the anode.

In the G-M tube, ionization proceeds until nearly all the gas atoms in the sensitive volume are involved, whereas in proportional counters, the amount of ionization produced per count is a function of the primary ionizing event as well as of the tube characteristics and operating conditions. The number of ion pairs produced per count can vary from one to 10^4 or even 10^6 when the incoming ionizing radiation produces as few as one primary ion pair.

As in the G-M tube, the primary electrons in the initial ionizing event produce secondary electrons by collisions with neutral gas molecules. As before, these events cascade to form a Townsend avalanche (see Figure 2.5). Since the voltage is lower in a proportional counter, this avalanche takes place very close to the anode and spreads out only over a very short distance from where it begins. The avalanche terminates when all the free electrons in this local vicinity have been swept to the anode. The number of secondary ionization events is kept proportional to the number of primary ion pairs formed by the tube conditions, even though the total number of ions can be multiplied by many powers of ten. Since the number of electrons swept to the anode is large, the change in anode voltage is large and hence the signal associated with this count is quite strong.

2.1.3.2 Physical Characteristics

Proportional counter chambers in all cases have a fine wire, usually of the order of 25 μm diameter, for the collector electrode. Since the Townsend avalanche is a local phenomenon with respect to position along the anode, in order to keep the multiplication proportionality, it is very important that the wire be extremely uniform in diameter along its length. A variation in anode characteristics would change the voltage characteristics in the vicinity of the variation and obviate strict proportionality along the anode length. Therefore, the anode is usually made of tungsten.

Proportional chambers are generally cylindrical in shape with the central wire as the anode and the tube walls forming the cathode. The sealed tube type of proportional counter is very similar to the G-M tube in construction (see Figure 2.1). However, it is necessary to introduce into the proportional counter some method of maintaining uniform voltage characteristics along the entire length of anode. Since this is difficult near the ends of the anode, the counter is generally designed so that events occurring near the ends do not undergo any gas multiplication. An abrupt transition then takes place between these "dead" regions and the remainder of the sensitive volume.

2.1.3.3 Filling Gas

The filling gas for proportional counters is usually one of the inert gases (generally argon, neon, or krypton), mixed with a small amount of polyatomic gas, such as methane or isobutane. The addition of the polyatomic gas makes the proportionality characteristics of the chamber more independent of operating voltage. A mixture of 90% argon and 10% methane, known as P-10 gas, is common, as is a mixture of 96% helium and 4% isobutane. A mixture of 1.3% isobutane and 98.7% helium, known as Geiger gas, lowers the operating voltage of the counter, but also changes the proportionality characteristics. A fill gas composed of 64.4% methane, 32.4% carbon dioxide, and 3.2% nitrogen approximates the composition of biological tissue. Proportional counters used for thermal neutron detection can be lined with boron or filled with boron trifluoride ($^{10}BF_3$) or ^3He. For fast neutron spectroscopy, the tube is filled with hydrogen, methane, helium, or some other low atomic number gas. The tube is generally operated at atmospheric pressure (101.3 kPa), but sealed tubes can be operated at pressures between 10 and 60 kPa.

2.1.3.4 Resolving Time

In a proportional counter, the resolving time is dependent on its application. If only detection is necessary, as opposed to energy measurement, the pulse time separation is much less important. Recall that in a proportional counter, the secondary ion production is a localized phenomenon with respect to the anode. Hence the tube can receive another pulse and amplify it as long as the new positive ion sheath is formed at another position along the anode. As the multiplication of ionization increases, ion production spreads out along the anode and the resolving time increases.

When only detection is required, a resolving time of 0.2 to 0.5 μs can be achieved. When energy is to be determined as well, then the resolving time can be as much as 100 μs. An important factor in determining the resolving time of the counter is, of course, the electronics that follow the detector and turn the change in anode voltage into a measurable quantity. Figure 2.6 shows some typical gas-filled chambers attached to a variety of readout components.

2.1.4 Windowless Gas Flow Counters

A common application of counting in both the proportional and Geiger region is to use the principle of gas flow. In this type of counter the sample is generally inserted directly into the detector, hence the designation, windowless. Figure 2.7 shows a windowless gas flowcounter in which the anode is placed on one side only of the sample. This is known as 2π geometry, because the detection volume essentially forms a hemisphere.

Figure 2.6: Typical gas filled detectors in different shapes and sizes. Also shown are different types of readout components to which these units can be attached. The chambers and readout components are interchangeable. (Courtesy of Fluke Biomedical.)

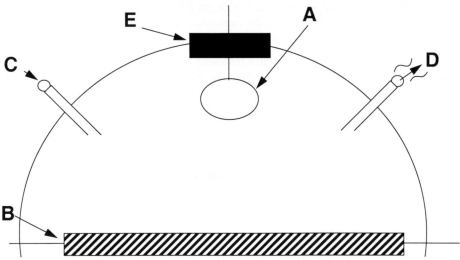

Figure 2.7: Example of a windowless gas-flow counter with 2π geometry. Note that the shape of the detector is mainly hemispherical. The anode is at **A** with the sample holder and sample at **B**. Gas is introduced through the opening **C** and ejected through **D**. The anode insulator is shown at **E**.

Another type of geometry known as 4π is shown in Figure 2.8. Here there are two anodes and hence two counting volumes that can be operated either independently or in coincidence. This geometry is especially useful for absolute counting of low-energy beta particles that may be scattered backward as noted in Chapter 1. When the sample is inserted, some air will enter the chamber. This is purged by a rapid flow of gas. While counting proceeds, the flow rate may be reduced to a very low level so that the gas consumption is nominal.

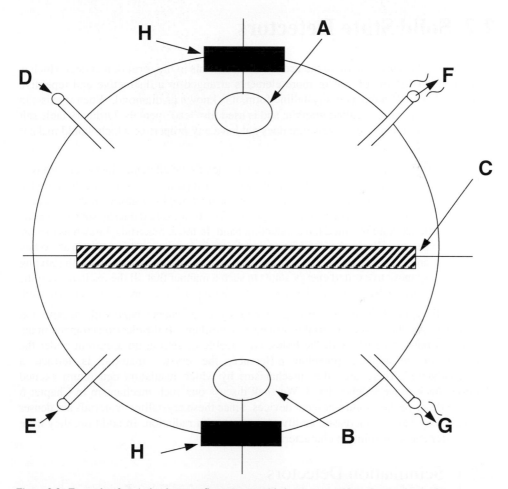

Figure 2.8: Example of a windowless gas-flow counter with 4π geometry. Note that the shape of the detector is roughly spherical. There are two anodes at **A** and **B**, with sample holder and sample at **C**. Gas can flow independently through both halves as pictured (**D,E,F,G**) or through both halves together. Each anode has a separate measuring apparatus. These can be operated independently or in coincidence. The anode insulators are shown at **H**.

Both mixtures of 90% argon with 10% methane and 96% helium with 4% isobutane are frequently used counter gases. The most important application of flow counters is for counting alpha particles, low-energy beta particles, and low-energy (soft) X-ray photons. For the latter two particles, energies as low as 1 keV have been detected. The upper limit on beta particle energies that can be counted by this method is 100 to 200 keV as determined by chamber size and gas pressure. Flow counters are always operated at atmospheric pressure.

2.2 Solid State Detectors

These types of detectors involve the use of solid materials in the form of a crystal which is an aggregate of atoms of one or more elements arranged in a distinctive and repetitive pattern. When carbon exists in crystalline form, it is known as diamond. When it exists in noncrystalline form, it is called graphite and is used for "lead" pencils. Ordinary table salt is a crystal, but its crystalline structure does not have any properties which would make it useful as a radiation detector.

Electrons in a crystal exist in distinct energy levels called bands, just as electrons in an atom exist in distinct energy levels called shells. In a crystal the electrons, especially the outer most or valence electrons oscillate between nearest neighbor atoms in the structure, giving some of these substances unique properties. The first excited energy state or "band" above the valence band is called the conduction band. In those materials, known as metals, electrons exist naturally in the conduction band, but in insulators this is not true. When electrons in an insulator gain energy, for example, as a result of interactions with ionizing radiation, they move to excited energy states in such a manner that all the excited electrons are in the conduction band and all the "holes" or missing electrons are in the valence band.

If the excited electrons return spontaneously to the valence band with the emission of light photons, the crystal material is termed a "scintillant". If the electrons trapped in the conduction band, together with the holes, are capable of setting up a current under the influence of an applied potential difference, the crystal material is termed a "semiconductor". There are other mechanisms by which insulators can return excited electrons back to the valence band. We shall discuss one such mechanism in Chapter 6 when we describe thermoluminescent devices. Since these crystalline materials are denser than gas, they have an increased ability to interact with radiation. In addition, they have much better energy resolution characteristics.

2.2.1 Scintillation Detectors

One of the oldest techniques for detecting ionizing radiation makes use of the scintillation light produced in certain organic and inorganic materials. Modern scintillation detectors were developed primarily in response to the need for detecting higher counting rates and consequently having shorter resolving times. In addition, this technique allows measurement of the energy spectrum of beta particles and photons particularly at relatively high-energy values.

A number of different materials are employed as scintillators. These include single solid organic crystals, organic materials dissolved in organic liquids or solids, plastics, thin films, single crystals of inorganic material usually doped with some impurity atoms, inorganic powders, glass, and high purity gases, usually noble gases, such as xenon and krypton.

It is important that all scintillation materials have a high efficiency for conversion of energy deposited by ionizing radiation to light (or very near ultraviolet) photons. This process is called fluorescence and indicates the prompt emission of visible light photons from a material following excitation of some sort. Phosphorescence, on the other hand, signifies that emission of a longer wavelength photon takes place and that the characteristic emission time is significantly delayed. Delayed fluorescence indicates that the photon emission energy is the same as in prompt fluorescence, but that the emission time following excitation is long. A good scintillator for radiation detection purposes should convert a large fraction of the incident energy to prompt fluorescence with a time delay of 10^{-8} s or less. The emission of energy through phosphorescence and delayed fluorescence should be minimal. In addition, it is necessary that scintillators be transparent to their own fluorescent radiation and that the photons emitted have energies, known as a spectral distribution, consistent with the response of available collecting apparati usually photocathodes. (Photocathodes play a role for scintillators similar to that of anodes which are the collectors for gas-filled detectors. However, there is no analogous attraction of the photons to the photocathode, but rather an optical coupling is required to channel the light. This is discussed later in this chapter.)

2.2.1.1 Principles of Operation

The process whereby a substance absorbs energy and then re-emits it as photons in the visible or near-visible energy range, is termed luminescence. The initial excitation energy for this process has many origins such as light chemical reactions, mechanical strains, heat, and ionizing radiation. The last named causes a scintillation because of the excitation or ionization produced in the substance. Only scintillants that de-excite promptly, i.e., through prompt fluorescence in 10^{-8} s or less, are useful as radiation detectors. However, most energy deposited in the substance by ionizing radiation is quickly degraded into heat. In choosing a scintillation material, therefore, it is important to determine the efficiency with which the kinetic energy of charged particles is converted into visible light and also to determine that this conversion is linear, i.e., that the light yield is proportional to deposited energy over as wide a range as possible.

To discuss the mechanism of the scintillation process, it is convenient to divide the materials used into the following five categories: organic crystals, liquid solutions of organic materials, solid solutions of organic materials, inorganic crystals, and noble gases.

Organic Crystal Scintillators. In all organic materials, the scintillation process is a molecular process. Practical organic scintillators have a certain electron symmetry property. This can best be understood in terms of an energy level diagram such as is given in Figure 2.9. The lowest level pictured is a singlet state (designated s_0) which represents the ground state energy of a particular electron. This ground state has associated with it a number of allowed molecular vibrational states (shown as dashed lines). An electron can be excited by ionizing radiation, to one of the higher singlet states (s_1, s_2, represented by the upward pointing arrows in the figure). Each one of these excited states is associated with one or more vibrational states (the solid lines in the figure indicate the singlet electronic states and the dashed lines represent vibrational states). The electron then decays by nonradioactive means back to the lowest nonvibrational level of the first excited state. It then decays by fluorescent radiation back to the ground state itself or to one of its vibrational levels. Two alternatives, other than prompt fluorescence are possible. First, it can happen that the electron transfers over to one of the lower lying triplet or other allowable electronic states. Then it has two choices, it can either return to the ground state by emittance of phosphorescent (different energy) photons, or at a later time, it can return to the original lowest level excited state and decay to the ground state emitting a delayed fluorescent photon. Either of these two mechanisms is undesirable. Secondly, the organic molecule can dissociate if the electron is excited to a high enough level. In either of the two cases above, the output of the scintillant is quenched, i.e., not realized. However, if the electron returns promptly to the ground or a vibrational ground state, the fluorescent photon it emits is indicative of an ionizing radiation event. Prompt fluorescence usually takes less than 10^{-8} s. It is also noteworthy that fluorescent photons are not energetic enough to excite an electron to an allowable energy state and thus are not re-absorbed by organic materials (which are thus transparent to their own radiation). The most widely used organic crystals are anthracene and transstilbene.

Organic Liquid Scintillators. In this type of detector, an organic scintillant is dissolved (used as a solute) in a liquid organic solvent. Sometimes a third substance is added to the solution. This third substance acts as a wavelength shifter, i.e., a fluorescent material that changes the photons produced by the main solute to somewhat longer wavelengths (lower energy) so as to more nearly match the response of the photocathode. Some substances used as solutes are p-terphenyl, 2–phenyl–5–(4–biphenyl)–oxazole (PBO), 2–phenyl–5–(4–biphenyl)–1,3,4–oxadiazole (PBD), and 1,4–di–[2–(5–phenyloxazolyl)]–benzene (POPOP). This last is also used in trace amounts as a waveshifter. Some good solvents are xylene, toluene, and phenylcycolhexane.

Scintillator efficiency increases rapidly as solute concentration initially increases, then passes a broad maximum before saturation is reached. As an example, maximum efficiency for p-terphenyl in toluene occurs for a solute concentration of about five grams per liter (5 g/L).

Ionizing radiation produces excitation in the solvent which is quickly transferred, before quenching, to the solute. The energy transfer mechanism is not well understood. However, the solute traps and degrades the excitation energy (as depicted in Figure 2.9) and subsequently, it is radiated as characteristic fluorescent radiation.

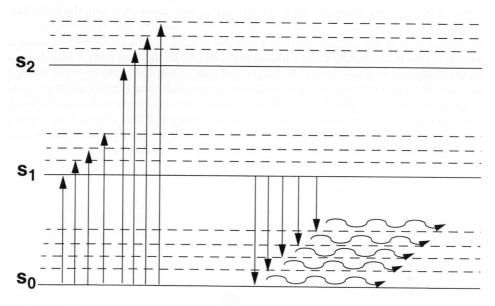

Figure 2.9: Electronic energy levels and molecular vibration states in a typical organic material. *Solid lines* indicate singlet electronic states. *Dashed lines* indicate vibrational energy levels. *Arrows pointing upward* show electronic excitations. *Arrows pointing downward* show fluorescent electronic transitions.

Liquid scintillators are widely used when the sources of ionizing radiation can be dissolved as part of the scintillator solution. In such a case, the efficiency is nearly 100% because all the radiation emitted by the source immediately passes through some portion of the detector. This a widely used technique for counting activity from such low-level beta emitters as ^{14}C or ^{3}H.

Organic Liquid Scintillators in Solid Solution. Organic scintillators can be dissolved in a solvent and the solution subsequently polymerized. Essentially, many kinds of plastic scintillation detectors can then be made. It is possible to produce the plastic as a thin film in order to detect even weakly penetrating particles such as heavy ions. Plastic scintillators can be made in a great variety of shapes and sizes. Although organic scintillants are generally useful for detecting fast electrons, alpha particles, and fast neutrons (through detection of the proton recoil process), their low atomic number constituents (H_2, C, and O_2) make them unsuitable for detecting photons in the X- and gamma-ray photon energy region. Plastic scintillators can have high atomic number elements such as lead or tin added. This increases their efficiency for low-energy photons, but decreases their light output. Thus, although they are cheaper than inorganic crystal scintillators, their energy resolution is considerably inferior.

Inorganic Crystal Scintillators. Inorganic scintillators are usually crystals of inorganic salts impregnated with small amounts of impurities as luminescent process activators. The mechanism for scintillation production will be discussed in terms of

Figure 2.10. In a pure inorganic solid crystal the electrons associated with the individual atoms and/or molecules exist in common energy states called "bands". Between the bands are energy regions that are forbidden. In a material suitable to be a scintillant, the valence band is normally completely filled with electrons, and the conduction band, which lies at a higher energy, is normally empty. If charged particles of sufficient energy move through the solid material, electrons can be moved from their normal state (known as lattice sites) to the conduction band. The electron vacancy, called a hole, will move to the valence band. Both the excited electron and the hole move independently and freely through the crystal. Sometimes, however, electron and hole stay bound together in a configuration known as an exciton. They are still free to drift through the crystal, but they do so as a unit.

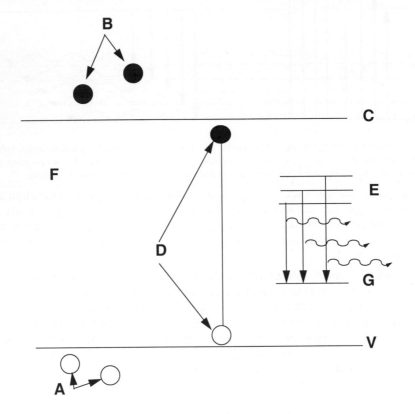

Figure 2.10:   Typical band structure of an inorganic solid. The line **V** marks the highest energy value contained in the filled valence band. The line **C** marks the lowest energy value contained in the conduction band. The energy states **F** between **V** and **C**, are ordinarily forbidden. **A** represents holes in the valence band caused by excitation of the electrons **B** to the conduction band. **D** represents an exciton pair drifting between the conduction and the filled valence band **V**. **G** denotes the ground state of the impurity used as an activator. When an impurity atom captures an exciton, or successively, a hole and an electron (order is not important), it is raised to one of its excited states **E**. If a transition is allowed between this excited state and the ground state, it occurs as shown and a light photon is emitted.

Any imperfection in the crystal, such as impurity atoms, will create energy levels in the forbidden band at isolated points. The excitons, holes, and free electrons produced by the ionizing radiation, drift through the crystal until captured by such a site. After the capture of an exciton or a hole and an electron in either order, the activator or impurity atom is raised to excited states. If a transition between the particular excited state and the ground state is allowed, it occurs in about 10^{-8} s with the emission of a light photon. If the excited state is a metastable one, the photon emission, if any, is slower. By a good choice of inorganic solid and activator then, most of the original energy deposited by the ionizing charged particle can be converted into light photons. The use of the activator or impurity atoms is essential to both shorten response time and to lower the energy of the emitted photon to the visible light region. (The natural energy difference between the valence and conduction band in the inorganic solid crystals used is usually higher than the visible light region.)

Some common solid inorganic scintillants are sodium iodide with thallium, NaI (TI), cesium iodide with thallium, Cs(TI), cesium fluoride, CsF, bismuth germanium oxide, ($Bi_4Ge_3O_2$ known as BGO), lutetium oxyorthosilicate with cesium, (Lu_2SiO_5:Ce known as LSO), and lithium glass. Because of their density, inorganic solid crystals are efficient for attentuating relatively high-energy photons. Inorganic scintillators range in size, but are limited at the upper end by the difficulty in producing very large single crystals.

Noble Gas Scintillators. High purity concentrations of the noble gases xenon and helium can be useful scintillators, for example, xenon is used in many computed tomography (CT) scanners. If the incident radiation passes through the gas and interacts, it leaves a population of excited gas molecules behind. As the excited molecules return to their ground state, through various mechanisms, both visible light and ultraviolet photons are emitted. Hence, either a light collector sensitive to ultraviolet must be used, or a small concentration of a second gas such as nitrogen must be used to wavelength shift the emission. A variety of parallel modes of de-excitation compete with photon emission, thus rendering the overall scintillation efficiency of gases characteristically low. However, the transition time is very short, typically, 10^{-9} s or less, making them attractive as fast radiation detectors. In addition to fast response, gas scintillators can be made in variable sizes and shapes and tend to be unusually linear over a wide range of energies. Their major disadvantage is low-light yield. Their application to nonpenetrating particles such as heavy-charged particles and fission fragments is also limited. (Note that the use of noble gases in gas-filled detectors was discussed earlier.)

2.2.1.2 Light Collection

A primary factor in a scintillation detector is the efficiency of light transfer from the point of origin in the luminescent material to the photocathode. Organic and inorganic solid crystal or plastic scintillators can be cut and shaped as appropriate, liquids must be placed in appropriate containers. Common shapes include solid circular cylinders, flat disks, and solid cylinders with recessed holes (known as "well counters").

The scintillant proper is encased for mechanical protection and control of ambient light. If it is hygroscopic, as NaI(TI) is, it must be hermetically sealed. For example, a sealed NaI(TI) unit might be surrounded by a layer of MgO or Al_2O_3 for light reflection and have a glass window on one face for transmission of light to the photocathode. Both casing (usually aluminum) and powdered reflector are minimized in thickness to prevent secondary electrons and degraded gamma-ray photons from forming in the casing. When the material is not hygroscopic, aluminum foil may be sufficient to prevent transmission of outside light. If low-energy betas are to be counted, a section of the main foil cover could be replaced by a much thinner aluminum window.

To aid in transmitting light between the scintillant and the photocathode, a material known as a light pipe is sometimes used. The light pipe conducts the light using internal reflection from entrance to exit. Lucite, quartz, and glass have been used as light pipes. A light pipe can help to prevent the trapping of light in the scintillator. It can also be used when the scintillator is placed at a distance from the photocathode or to spread the light over a large area of the cathode face.

2.2.1.3 Resolving Time

Although the scintillations themselves occur very quickly, about 10^{-8} s, the measuring apparatus associated with these detectors is usually not as fast. The resolving time associated with conventional scintillation detectors is typically $1 \cdot 10^{-6}$ to $5 \cdot 10^{-6}$ s. Special circuits have been developed with resolving times as low as $2 \cdot 10^{-7}$ s. Some organic scintillators, particularly stilbene, which has a decay time of $5 \cdot 10^{-9}$ s, have been used successfully with a measuring apparatus having a resolving time of $1 \cdot 10^{-7}$ s.

2.2.1.4 Remarks

Scintillants are widely used in applications other than as radiation detectors. For example, scintillators are used as the active element in film-screen systems in conventional radiology X-ray units. Photocathodes as collecting devices for light output are also commonplace. For example, they are the active element in image intensifiers, vidicon, and plumbicon tubes, and many others. In radiation detectors, the photocathode is generally incorporated into a device known as a photomultiplier tube. In essence, the light photons impinge upon the photocathode and release electrons from it.

The number of electrons released is related to the energy of the light photon. This original number of electrons is multiplied by a fixed proportionality constant in the photomultiplier tube. (This is similar to the production of secondary, tertiary, ..., electrons in the proportional counter.) The mechanism of multiplication is accomplished in a variety of ways which are specific to each type of photomultiplier tube. The electrons produced are collected by the anode of the photomultiplier tube and the resulting voltage change is recorded and analyzed by the measuring apparatus. Photomultiplier tubes come in many shapes and sizes and employ many different electron multiplication techniques. Consult the references cited in the bibliography for details on specific photomultiplier tubes.

2.2.2 Semiconductor Detectors

Another type of solid state detector, known as a "semiconductor" detector, also takes advantage of the properties of particular crystals. In this case, the crystal is a very good insulator which under certain conditions, can be made to conduct electricity. The typical semiconductor detector can be thought of as a sandwich with the crystal in the center and a positive and negative electrode attached to each side. Under ordinary circumstances, all the electrons in the crystal structure are in the lowest allowable energy states. However, when energy is deposited in the crystal, the electrons are excited to energy states near the valence band, as in the case of the scintillator, but in a semiconductor the electrons get trapped in this higher energy state. As usual, all the missing electron vacancies or holes propagate to the valence band. When a potential difference is applied to the crystal, the electrons travel toward the positive electrode and the holes toward the negative electrode, thus generating a current. It is important that current not flow when there is no applied potential and it important that all the charge carriers be detected when there is an applied potential. Semiconductor detectors are similar to ionization chambers in that the energy deposited by the incoming ionizing radiation is exactly equivalent to the number of electron–hole pairs produced, thus it is important that all the charge is collected. As with an ionization chamber detector, semiconductor detectors suffer from the problem of recombination of the produced charge. The ability of these detectors to furnish accurate energy data is degraded by the trapping of electrons which creates within the crystal structure a space charge that reduces the electric potential between the electrodes, further facilitating recombination of electron hole pairs.

2.2.2.1 Principles of Operation

An electron–hole pair is produced when incoming ionizing radiation deposits energy which allows an electron to move to an excited or higher energy state. In certain crystal materials, the electrons get trapped in an energy state, known as the conduction band, and cannot decay back to a lower energy state. Semiconductor materials have the property that, under the influence of an electric potential difference, the electrons in the conduction band can move and at the same time the crystal structure can propagate, between neighboring lattice points, a net positive charge because of the missing electron. This movement of charge carriers generates a current flow, hence the term "semiconductor". A charged particle which interacts with a solid loses energy by transferring it to electrons in the medium. This energy is used to "excite" these electrons to a higher allowable energy state, just as in the case of the scintillant. The number of electron–hole pairs produced is given by:

$$\text{Number of Pairs} = \frac{E}{\varepsilon} \qquad (2.1)$$

where E is the total energy deposited by the incoming ionizing radiation and ε is the average energy to produce an electron–hole pair. This interaction can be characterized in a non-relativistic context by:

$$E_{\text{max transferred}} = \frac{4\,m \cdot E}{M} \qquad (2.2)$$

where E_{max} is the maximum total energy deposited by the incoming charged particle and m and M are the mass of the electron and the incoming charged particle respectively.

For a 4 MeV alpha particle, for example, E_{max} is about 2 keV which is much more energy than is needed to raise one electron from the valence or deeper lying energy state to the conduction band or a higher energy state. This results in the crystal being in a highly excited state for about 10^{-12} s and then the holes are propagated to the top of the valence band and the electrons to the bottom of the conduction band. Some typical values for ε are 3.23 eV for silicon (Si) and 2.84 eV for germanium (Ge). Thus we see that the energy needed to create an electron–hole pair is about 30 eV less than the energy needed to ionize an air molecule in an ionization chamber (33.85 eV). This leads to improved counting statistics and energy resolution. The values of ε quoted are actually almost three times that needed to raise an electron to the conduction band. The remaining energy is dissipated through the strong coupling of the electron to the crystal lattice and through vibrations of the solid. Because ε is independent of the energy and mass of the primary ionizing particle, semiconductor detectors are useful in many different applications. Some of these include heavy particle detection, beta spectrometry, particle identification, measurement under high–background conditions, and even the ability to replace the photomultiplier tube in scintillation detectors (thus producing a higher signal–to–noise ratio).

Electron–Hole Mobility. When an electron–hole pair are produced in the presence of an applied potential difference, each particle moves toward its respective electrode; i.e., the electron moves toward the (relatively) positive electrode. The charge q induced by the motion of a charge Q through a distance Δx is given by:

$$q = Q\frac{\Delta V}{V_a} \tag{2.3}$$

where V_a is the total potential difference applied across the electrodes and ΔV is the potential difference corresponding to Δx when no space charge is trapped between the electrodes. The time it takes for an electron or hole to travel through the material from its point of production to the electrode is called the transit time. This time is determined by the "mobility" or number of centimeters per second per volt per centimeter that the particle can travel through the material. As an example, consider a homogeneous detector made from silicon. A typical electron mobility value is 1,500 cm²/V·s whereas a typical hole mobility value is 500 cm²/V·s.

Space Charge. For good energy resolution, it is important to collect all the electron–hole pairs. Sometimes electrons get trapped in an imperfection in the lattice structure of the crystal and do not reach the electrode. When this happens, two effects are observed. The first is a polarization effect where the trapped electrons set up a potential difference which opposes the applied voltage. The second is the production of secondary currents because of the dielectric-relaxation time. If this time is less than the transit time of the electron in the material and if the material allows this phenomenon, secondary currents will flow. The efficiency of charge collection may be written as:

$$\eta = \frac{q}{Ne} \tag{2.4}$$

where q is the charge on the collecting electrode, N is the number of electron–hole pairs released (E/ε), and e is the charge on the electron. By relating the charge collected to the fraction of the distance traveled through the detector, one can relate the efficiency to the history of charge carriers as they attempt to cross the detector from the point of production to the electrode. The two mechanisms which prevent the charges from reaching the electrode are recombination and the build up of space charge as a result of electron trapping.

Material Requirements. The properties that are important for semiconductor detectors include the charge-collection characteristics and the resistivity of the material. A sufficiently high potential difference must be applied to the detector material to collect the charged particles produced by the incoming ionizing radiation, therefore, it is necessary that the material have a high enough resistivity so that charge will not flow in the absence of such interactions. Because the crystal structure of most materials is not perfect, there is always a small amount of leakage current present. Since a typical signal current is 10^{-6} A, it is important that the leakage current be significantly less so that it is possible to discriminate against it. Typical semiconductor detector materials should then have the properties of high resistivity (low leakage current), high mobility, and long carrier life-times for electrons and holes to facilitate collection of all the charge produced, and be capable of sustaining large potential differences between the electrodes. The second and third requirements together imply that the charge-carrier trapping lengths be much greater than the thickness of the collection region. Of all the possible materials available, silicon, germanium, and gallium-arsenide (most recently enhanced with indium) have the best material properties and can be made with a high degree of purity and crystal perfection, but the resistivity of the first two is too low. For this reason, homogeneous detectors are rare. Semiconductor detectors usually exist in the form of junction detectors which will be discussed below.

2.2.2.2 Types of Junction Semiconductors Detectors

The material requirements for semiconductors as outlined above which make homogeneous semiconductor detectors impracticable, can be overcome by introducing other (impurity) atoms into the original material and thus creating regions within the detector which are donor-rich (i.e., have more electrons present — often known as n-type regions) and other regions which are acceptor-rich (i.e., have more "holes" or are electron deficient — often known as p-type regions). The place where n-type material interfaces with p-type material is known as a "junction", and these types of detectors are often called junction detectors.

There are typically three different ways to produce junction detectors: diffusion, surface barrier emplacement, and lithium drifting. The detectors are known by the method used to produce them. The charge-deficient region that borders on the interface between the n and p type regions, is often termed the depletion or space charge region which actually forms the region of the detector sensitive to incoming ionization radiation. When this region is extended throughout the entire volume of the detector, it is said to be "totally depleted". Each of the three types of junction detectors is capable of being configured in this manner.

Diffused and Surface-Barrier Junction Detectors. By injecting a donor-rich impurity such as phosphorus to a shallow depth in originally high purity p-type silicon, a small n-type region can be formed near the surface of the material. In the n-type region thus created, the majority charge carriers are electrons in the conduction band, coming mostly from impurity atoms with electron energy states near the conduction band. In the original p-type region, the majority charge carriers are holes formed mostly by the excitation of the valence electrons to the electron acceptor states near the valence band. When the two types of materials are placed close together, the electrons tend to diffuse from the n-type material toward the p-type material and the holes do the reverse. This causes a depletion region to be formed by unfilled holes migrating to the volume near the conduction band in numbers greater than the number of electrons in the conduction band, and electrons migrating to the valence band also in numbers greater than the number of holes to be filled. This imbalance of charges within the depletion region sets up a small potential difference across the region which can be strengthened by the application of an external potential difference sometimes referred to as the "reverse bias" voltage. It is this depletion region, which is a region of high resistivity, that then serves as the practical sensitive volume of the detector for incoming ionizing radiation. (The sensitive volume can actually be extended a very small amount from the depletion region, but for practical purposes the two volumes are equal.) The diffused junction detector described here is known as n-p type. Similarly, one can start with high purity n-type silicon and diffuse into it acceptor-rich or p-type impurity atoms such as those from gallium or boron. This then forms a p-n type diffused junction detector. Figure 2.11 gives a schematic representation of a diffused n^+-p junction detector and Figure 2.12 depicts a surface-barrier junction detector..

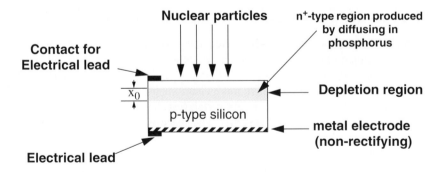

Figure 2.11: A schematic representation of a diffused n^+-p junction detector.

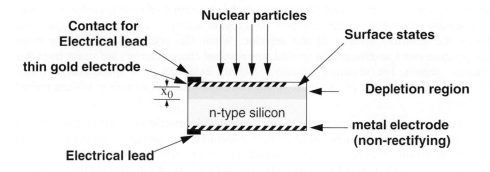

Figure 2.12: A schematic representation of a surface-barrier junction detector.

A depletion region of high resistivity similar to the one described above, can also be formed on the surface of materials. If a high-purity n-type silicon crystal is chemically etched, surface states will form by spontaneous oxidation. A high density of holes is then induced by these states which in essence becomes a p-type layer. Evaporation of a thin layer of gold makes electrical contact with this layer possible and forms a barrier to the flow of current when a potential difference is properly applied across the n-p junction. The upper limit on the thickness of the depletion region in a diffused or surface-barrier junction detector is determined by the maximum resistivity of the material (usually silicon) available and the maximum potential difference in the form of the reverse bias that can be applied.

Surface-barrier junction detectors can be produced in a large number of shapes and sizes in the laboratory. However, diffused detectors are better adapted to quantity production of similar units. If the other properties of the detector are equal, the surface-barrier detector usually has better energy resolution. If proper encapsulation is provided, there is no appreciable difference in the long-term stability of these two types of detectors, however, it is possible to expose only the surface-barrier type to air for prolonged periods of time with no deleterious effects.

Totally Depleted Junction Detectors. In the detector types described above, two regions are insensitive to incoming ionizing radiation because it is not possible to collect charge there. The first such region is termed the "window" and is that portion of the detector through which external radiation must travel to the depletion or intrinsic region. The second region is the undepleted region beyond the space-charge region in the diffused or surface-barrier junction detector or the uncompensated region in lithium drifted detectors. One method of eliminating these regions in diffused or surface-barrier detectors is to raise the product of the resistivity and the applied potential difference so high that the depletion region extends throughout the entire thickness of the detector. Similarly, lithium drifted detectors can be constructed so that the intrinsic region extends to the rear electrode.

Lithium-Ion Junction Detectors. When lithium, which is a donor impurity, is diffused into one surface of a p-type crystal such as silicon, and a reverse bias along with a temperature increase is applied for a short time, the junction between the donor region (n) and the acceptor region (p) can be lengthened. In this case the donor and acceptor impurities compensate each other in equal numbers. This region is known as the intrinsic region and forms the sensitive volume of the detector. When the crystal is returned to room temperature and a reverse bias is applied, any residual charge carriers are swept from this intrinsic region. The sensitive volume is slightly extended by a space charge which develops at either edge of the intrinsic region. Figure 2.13 illustrates a lithium drifted detector.

One method of constructing a totally depleted detector is to allow the diffused phosphorus or lithium to form a high intensity n region at one surface of the detector. Then aluminum or boron may be used to form a high intensity p region at the second surface. The advantages of this type of detector are that either surface may now be used as the "window" for entrance into the detector and hence the best side can be chosen. In a totally depleted detector there is no series resistance present between the sensitive region of the detector and the electrodes, thus leading to improved mobility of the charge carriers. Figure 2.14 illustrates the principles involved in a totally depleted detector.

Depletion Region Thickness and Electrical Properties. For diffused and surface-barrier junction detectors, the thickness x_0 of the depletion region can be expressed as:

$$x_0 = [K\rho(V_a + V_0)]^{1/2} \tag{2.5}$$

Here, x_0 is expressed in centimeters, ρ in ohm-centimeters, V_a, the applied (reverse bias) voltage and V_0 the natural potential difference in volts and K is a constant depending on the material used and the mobility of the electrons and the holes. For a diffused-junction p-type silicon device, $K = 1.1 \cdot 10^{-9}$ cm/(V·ohm) and for a surface-junction n-type silicon device, $K = 3.2 \cdot 10^{-9}$ cm/(V·ohm) where the hole mobility is taken as 500 cm^2/(V·s) and the electron mobility as 1,500 cm^2/(V·s). From this one can calculate as a first approximation, the electrical properties of the detector in terms of the capacitance per unit area, C/A as:

$$\frac{C}{A} = \frac{1.061}{x_0} \tag{2.6}$$

The value of the capacitance coupled with the value of the applied (bias) voltage determine the maximum energy ionizing radiation particle of a particular type that can be detected.

For a lithium-ion drifted detector, both the thickness of the sensitive region and the capacitance of the detector are approximately independent of the value of the imposed bias potential difference. Detectors with relatively thick (1 cm) sensitive volumes can thus be produced so that higher-energy particles can be stopped; however, the resolution goes down as the transit times for the charge carriers increases due to detector thickness.

In a totally depleted detector, even in the case of a diffused or surface-barrier junction, the capacitance is now independent of the applied (bias) potential difference. Hence a voltage sensitive instead of a charge sensitive pre-amplifier (see below) can be used with this device, thus increasing the signal–to–noise ratio of the system.

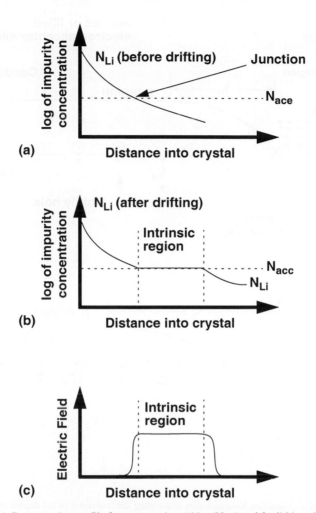

Figure 2.13: (a) Concentration profile for acceptor impurities (N_{acc}) and for lithium donor impurities (N_{Li}) in a diffused junction diode. (b) Concentration profile after the drifting process. (c) Electric field distribution with applied reverse bias.

2.2.2.3 Operating Characteristics of Junction Detectors

Like all other radiation detectors, semiconductor detectors must be integrated into a system (usually electronic) to report the measurement that is obtained. A semiconductor detection system usually has, as the first element in the read-out chain, a pre-amplifier which boosts the incoming signal.

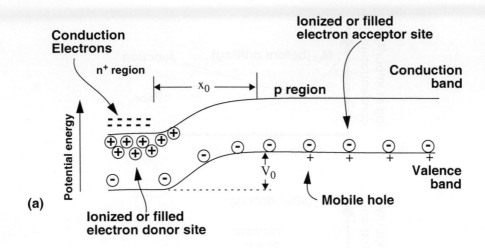

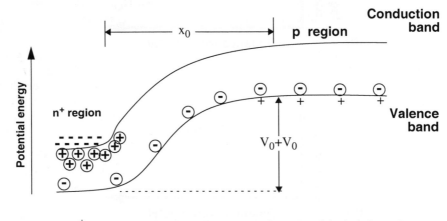

Figure 2.14: Diffused n^+-p junction band structure illustrating the formation of the depletion region. (a) No bias is applied. (b) A reverse bias is applied.

Energy Resolution — Pulse Height. For radiation detection, two types of pre-amplifiers are commonly used: those which are voltage sensitive (VS) and those which are current sensitive (CS). Although voltage sensitive pre-amplifiers are commonly used for ionization chambers and proportional counters, they have the disadvantage that the output voltage is a function of the induced charge divided by the total input capacitance (defined as the sum of the detector capacitance plus any other capacitance generated by the input circuit) of the pre-amplifier. While it is true that voltage sensitive devices have a faster resolving (rise) time and a slightly higher signal to noise ratio, this is not enough to overcome the fact that the output pulse height is dependent on the detector capacitance and the pulse decay time is dependent on the detector leakage resistance. Any variation in the detector capacitance or resistance thus changes the input capacitance or resistance respectively, of the pre-amplifier. In a current sensitive pre-amplifier, however, the output voltage is proportional to the induced charge and independent of the capacitance of the detector. It depends on the capacitance of the external circuit "feedback" loop which can be as low as a few picofarads. Both lithium drifted and totally depleted junction detectors have relatively constant, but unfortunately high capacitance, so although voltage sensitive devices might be acceptable for these classes of detectors, the signal is not sufficiently larger than the noise. Thus charge sensitive devices are the choice for these detectors also.

A further argument for the use of charge sensitive pre-amplifier devices is illustrated by the fact that the pulse height is not affected by the applied (bias) potential difference as in the case of voltage sensitive devices. The relative pulse height versus particle energy for different types of particles falls on a straight line for charge sensitive devices. This linearity is one of the main features that makes semiconductor detectors very useful instruments for determining the energy spectrum associated with a beam of ionizing radiation. Any departure from linearity for high energy particles implies that the range of the primary ionizing radiation has exceeded the thickness of the sensitive region of the detector. These maxima in pulse height versus particle energy are caused by the specific ionization properties of different types of ionizing radiation and may be used to distinguish between them. It is this property of semiconductors that makes them extremely valuable as radiation detectors.

The ability of semiconductor detectors to resolve different particle energies and types, rivals that of magnetic spectrometers at a much reduced cost and a much faster output rate. The factors that affect resolution can be categorized as the statistics of electron–hole formation, the noise generated in the detector and following electronics and some miscellaneous other effects. Of all of these, detector noise is usually the major contributing factor.

Window Thickness. The length of material through which a charged particle must pass to enter the sensitive volume of the detector is known as the window thickness. Any energy degradation that occurs in this region is lost from the output signal. The window thickness can be determined by measuring the pulse height produced by a collimated ionizing radiation source incident upon the detector at two different angles. For semiconductor detectors it is possible to make very thin windows so the energy lost in them can be neglected. In lithium drifted detectors, the window thickness can become very large if the lithium remains on the surface, so it is particularly important that good drifting techniques be used to construct these instruments.

Detector Area. This is the geometric property of the detector which determines the portion of the incoming ionizing radiation field that is intercepted by the detector. In semiconductor materials, increasing the detector area also has the effect of increasing the leakage current which in turn decreases the detector resolution. A large single crystal detector is also hard to produce. The detector area can be increased without the unwanted side-effects by grouping several smaller detectors together.

Leakage Current. That current which flows in a detector even in the absence of interacting radiation is known as the leakage current. Since this current is not part of the signal, but is part of the noise, it reduces the signal to noise ratio of the detector. There are two types of leakage currents: those which flow on the detector surface and those which flow within the detector volume. The first is determined by the detector construction, by the surface treatment, by the amount of accumulated contamination to which this surface leakage current is very sensitive, and by the ambient conditions of temperature and pressure. The second is determined by the bulk material properties which effect the diffusion of charge carriers into the depletion region (usually negligibly small), and the generation of charge carriers at the trapping centers in the depletion region. This latter must be controlled, else the detector becomes useless.

Resolving Time. This is the time during which the signal rises to the maximum pulse height. It is affected by the collection time for an electron–hole pair liberated by the radiation, the effective time constant of the detector and pre-amplifier combination and the characteristics of the following electronics which complete the signal generation. Of these, the first is the most critical and the most difficult to minimize.

The most important factor involved in the electron–hole pair collection or transit time is the exact location of production. If all the electron–hole pairs are produced in the sensitive volume, then the collection time is much shorter than if some pairs are produced outside this region and must diffuse into it. In the lithium drifted and totally depleted detectors, as in the homogeneous detectors, the potential difference is constant throughout the sensitive volume provided the applied voltage is high enough, and hence the maximum transit time required for a charge carrier to cross the detector is:

$$t = \frac{d^2}{\mu V_a} \tag{2.7}$$

where d is the thickness of the sensitive volume, μ is the mobility and V_a is the applied voltage. In the diffused and surface-barrier junction detectors the potential difference as felt by the charge carriers is a function of position within the sensitive region and so a simple relation for calculating the transit time such as that given above does not apply. For these types of detectors, the collection time is proportional to ρ/μ where ρ is the resistivity, the mobility μ is assumed constant, and the potential difference follows a parabolic distribution in the sensitive volume.

Another problem encountered with silicon is that mobility of the charges starts to decrease as the applied potential difference is increased, until a limiting velocity of about 10^7 cm/s is reached. Thick sensitive volumes also increase the collection time. The transit time, however, is only one part of the chain. The time for the information to get to the pre-amplifier depends upon the time to transmit the data through the undepleted region of the detector and through any existing contact resistance. When all these factors have been taken into account, resolving times for semiconductor detectors most often fall in the range of 10 ns to 1 μs.

Irradiation Effects. The integrity of semiconductor detectors is compromised by the type and intensity of the radiation to which they are exposed. For example, for 5.5 MeV alpha particles, a flux of 10^8 particles/cm^2 causes an increase in the reverse current and a flux of $2 \cdot 10^9$ particles/cm^2 causes a deterioration in the resolution. Likewise for a flux of fast neutrons exceeding 10^{11} particles/cm^2, detector properties are observed to deteriorate.

When semi-conductors are irradiated extensively with particle intensities beyond their tolerance, lattice defects and other imperfections result. These imperfections serve as trapping centers which lower the charge-carrier life-time and increase the rate of charge-carrier generation from the trapping centers. The resistivity of the material also increases. These effects all contribute to an increase in resolving time, and a decrease in charge particle collection efficiency, in resolution, and in homogeneity of the sensitive volume.

Remarks. Semiconductor detectors are used for a wide variety of purposes, ranging from X-ray astronomy to charge-coupled devices (CCD). Silicon and germanium are still widely used, although indium-gallium-arsenide (InGaAs) is also popular as it needs no cooling when in use and provides a resolution very close to that of germanium. The choice of detector material is largely dependent on the application. The energy resolution of semiconductor detectors is superior to that of scintillants, but the expense and fragility of both of these types of detectors make them less attractive as radiation protection devices than gas-filled devices. Nevertheless, when energy resolution is important, semiconductors are the detectors of choice.

2.3 Review Questions

1. Of what two parts does every radiation detection system consist?

2. How are different types of detector systems characterized? Name some examples of each.

3. What are two common modes of operation for any radiation detector?

4. What are the three common types of gas-filled detectors?

5. Briefly describe how electrons, once liberated in gas filled detectors, travel toward the anode.

6. Describe what happens in the vicinity of the anode in an ionization chamber.

7. Describe what happens in the vicinity of the anode in a G-M tube.

8. Describe what happens in the vicinity of the anode in a proportional counter.

9. Describe the structure of a crystals, in particular describe how the electrons exist in the crystal structure.

10. What is the difference between a conductor and an insulator?

11. Are there many ways in which electrons in the conduction band of an insulator can decay back to the valence state?

12. What is meant by scintillation? What is the difference between fluorescence, phosphorescence and delayed fluorescence?

13. What plays the role of the collector in scintillation detectors?

14. What is meant by a semiconductor? What is the difference between homogeneous and junction detectors?

15. How many types of junction detectors are there? What makes them different from each other?

16. What are some of the advantages of semiconductor detectors?

2.4 Problems

1. An ionization chamber is filled with air at atmospheric pressure (101.3 kPa). If it takes 33.85 eV of energy to create one ion pair in air, how many ion pairs are created by a 500 keV beta particle which deposits all its energy in the chamber? How many electrons from this event are collected at the anode (provided there is no recombination in the chamber)?

2. A proportional counter is filled with argon gas at atmospheric pressure. The energy necessary to ionize Ar is about 26 eV. If the multiplication factor is 10^3, how many ion pairs are created by a 500 keV beta particle which deposits all its energy in the chamber?

3. A G-M tube is filled with helium at a pressure of 9 kPa. The energy necessary to ionize He is 32.5 eV. If a 200 keV photon enters the chamber and through a Compton interaction releases a 50 eV electron, will a count be observed (provided the electron deposits its energy within the sensitive volume of the detector)?

4. If the resolving time of G-M tube is 200 μs and the photon beam contains 4,000,000 photons per second, how many counts will the G-M tube register? What is the efficiency of the counter?

5. A solid organic scintillator has an overall resolving time of $5 \cdot 10^{-7}$ s. How many counts per second can this detector register under ideal conditions?

6. A silicon-based lithium drifted junction detector has an electron mobility of 1,500 cm^2/V·s and a hole mobility of 500 cm^2/V·s. If an electron–hole pair is produced in the middle of the sensitive volume, exactly 5 cm from each collecting electrode, what will be the transit time of each particle? The applied voltage is 1.0 V.

7. The maximum transit time for a charge carrier to cross a particular semiconductor is 3 ns. The charge carrier mobility is 1500 cm/V·s and the applied voltage is 0.5 V. What is the distance the charge carrier must travel?

Chapter 3

UNITS ASSOCIATED WITH RADIATION PROTECTION

When living cells are exposed to ionizing radiation, they may absorb some or all of the energy carried by the ionizing radiation. It is the energy transferred from ionizing radiation to living cells which damages them. The amount of energy actually deposited by radiation in cells is known as absorbed dose. In terms of human biologic tissue damage, the factors that must be considered are the type of radiation encountered, the amount of energy transferred by that radiation and the radiosensitivity of the particular tissue or organ in which that energy is deposited.

Since it is not possible to routinely measure the amount of energy specifically absorbed by particular tissues or organs, we estimate that value by monitoring the surroundings or evaluating the amount and quality of the radiation to which a person is exposed. In this chapter, we study the physical units associated with these measurements in addition to the concepts used to translate physical measurements into estimates of biologic damage.

3.1 Exposure

When a beam of photons passes through matter, not all of the photons interact. For low-energy photons of the type encountered by most radiation workers, a special unit of exposure has been defined. This unit is based on the ionization of air, i.e., is quantified by the amount of charge of one sign released by photons as they pass through a specified mass of air. In general, one unit of exposure is designated as one coulomb (C), of positive or negative charge, released in one kilogram of air:

$$1 \text{ exposure unit } = 1 \text{ C} \cdot \text{kg}^{-1} \text{ (air)} \tag{3.1}$$

This equation reflects the way the unit of exposure is defined in the International System of Units (SI) which will be used throughout this text. Appendix A gives a history of units and a summary of units which are currently in use. In the former system of units (Appendix B), one exposure unit was defined as the roentgen (R), where $1 \text{ R} = 2.58 \cdot 10^{-4} \text{ C} \cdot \text{kg}^{-1}$ (air).

In order to measure exposure, the amount of ionization is determined using a special detector known as a standard ionization chamber. This chamber is used in a manner analogous to any international standard, such as those described in Appendix A. The standard ion chamber serves as a primary standard against which secondary standard ion chambers are calibrated. As can be seen from Figure 3.1, the standard ion chamber is composed of a unit cell of air with electrodes placed on two sides (top and bottom in diagram). Surrounding the unit cell are similar air-filled cells, but without electrodes in them. Consider a beam of photons moving from left to right, as shown in Figure 3.1. Photons pass through the first cell on the left, ionizing some air molecules along their path and thus producing electrons. As long as these electrons do not enter the central cell, they cannot reach the positive electrode.

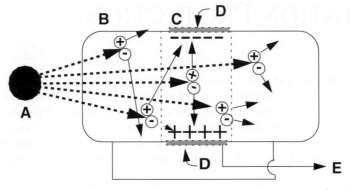

Figure 3.1: Outline of a Standard Ion Chamber. The photons are emitted by the source **A** and enter the outer cell **B**, which is grounded. The inner cell **C** is surrounded on both sides by air-filled grounded cells. Ions are collected on the electrodes **D**, which are connected **E** to a charge sensitive measuring device.

However, some electrons formed outside the central cell do enter it and do reach the positive electrode. These electrons give a false exposure value (they add to the number of coulombs collected on the central electrode, which ideally collects only electrons produced in the volume of the central cell) unless there is some way to compensate for them. At the same time, some of the electrons produced by ionization in the central cell pass into the cell to the right before they can be collected. It would be convenient if the number of electrons entering the central cell from the left were equal to the number of electrons leaving to the right. Such a situation is known as electron equilibrium (Figure 3.2).

The determination of exposure values using the standard ion chamber is valid only when electron equilibrium can be established. Unfortunately, since the density of air is small, higher energy photons do not readily interact with it thus making it hard to reach electron equilibrium. In fact, it is only practicable to define exposure values for photon energies up to about 3 MeV. Generally speaking, exposure units are an accurate measure only for photon energies between 1 keV and 1000 keV (1 MeV). It is important to remember that the exposure unit is defined only for photons in a fixed energy range interacting with air. This restriction limits the usefulness of the exposure unit considerably, therefore, additional units are defined for all kinds of ionizing radiation.

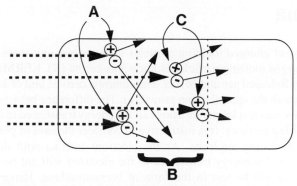

Figure 3.2: Electron Equilibrium. The electrons **A**, although produced outside the standard cell **B**, enter it. The electrons **C**, although produced in the standard cell, leave it. For electron equilibrium to be established, the number of electrons **A** must be equal to the number of electrons **C**.

3.2 Absorbed Dose

The path of ionizing radiation through matter is marked by a transfer of energy from radiation to the material through which it passes. This transfer of energy is typical of all kinds of ionizing radiation and not restricted just to photons. The amount of energy transferred is known as the absorbed energy, or the absorbed dose. The common derived unit of absorbed energy, or absorbed dose, is called a gray[1] (Gy) which is defined as one joule (J) of energy absorbed from any ionizing radiation in one kilogram (kg) of any material. In symbolic form:

$$1\,Gy = 1\,J \cdot kg^{-1} \tag{3.2}$$

The former unit of absorbed dose, known as the rad, is discussed in Appendix B. The relation between rad and gray is very simple:

$$1\,rad = 1 \cdot 10^{-2}\,Gy = 1\,cGy \tag{3.3}$$

Therefore, any absorbed dose values expressed in rad can be expressed directly in centigray (cGy). Many older tables of absorbed dose are expressed in terms of rad.

Frequently it is important to know not only the total absorbed dose, but also the time period over which that absorbed dose was acquired. This is expressed as the absorbed dose divided by the time period and is known as the absorbed dose rate. The common derived unit of absorbed dose rate is gray per second (Gy·s^{-1}).

---

1. The gray is named for the English radiobiologist Louis Harold Gray (1905–1965), who was instrumental in establishing experimental techniques to measure absorbed dose.

3.3 Kerma

The kinetic energy of *charged* ionizing particles liberated per unit mass of the specified material by *uncharged* ionizing particles is known as kerma (K). KERMA is an acronym for Kinetic Energy Released per unit MAss. The common derived unit of kerma is joule per kilogram $(J \cdot kg^{-1})$ with the special name gray (Gy). The difference between absorbed dose and kerma lies in the fact that kerma measures all the energy transferred to charged particles by uncharged ionizing particles. It is instructive to consider the case of photons interacting with matter and producing electrons. As the electrons interact with the material, it is possible that some of the energy transferred to the electrons will not be deposited in the material, but instead will be lost in the form of bremsstrahlung. Hence, the distinction between absorbed dose and kerma may be expressed as follows: absorbed dose is the energy per unit mass retained or absorbed along the path of the charged particle(s), while kerma is the energy per unit mass transferred to the charged particle(s). It must also be kept in mind that energy may be released (transferred) at one location (the interaction location), but not absorbed at this location. Thus energy may be released in a particular volume element and absorbed in it, or it may be released in one volume element and absorbed in a different one. Hence, absorbed dose and kerma may differ in value depending on the particular location in question.

The ICRP has established that it is acceptable to use such terms as "tissue kerma in air" or "tissue kerma in bone" because kerma, unlike exposure, can be quoted for any specified material at a point in free space or in an absorbing medium. Because over a wide range of photon energies, air kerma and tissue kerma differ by less than 10%, for radiation protection purposes, they may be considered equal in magnitude. Air kerma means air kerma in air. As kerma is independent of the complexities of geometry of the irradiated mass element and permits specification for photons or neutrons in free space or in an absorbing material, it has wider applicability than exposure. Kerma is generally determined with a radiation measuring instrument designed to approximate charged particle equilibrium conditions.

When this is done, then the value of kerma and the value of the absorbed dose are equal when measured in the same units, provided no significant amount of energy is lost. It is very convenient, dosimetrically, to link air kerma to exposure. Indeed it is found that for low-energy photons of approximately 100 keV, for example, which produce an exposure of one roentgen at a point in air, the air kerma, is about 0.87 cGy (0.87 rad; see Appendix B for the conversion from roentgen to rad; recall that 1 rad = 1 cGy) and the tissue kerma is about 0.95 cGy (0.95 rad). For a medium other than air, kerma depends upon the energy of the photons and the atomic composition of the medium.

Just as an absorbed dose rate was defined in the previous section, we can define a kerma rate and an exposure rate. Again the quantity is divided by the total time period during which the irradiation occurred. The common name for kerma rate is gray per second $(Gy \cdot s^{-1})$; the common unit of exposure rate is coulomb per kilogram·second $(C \cdot (kg \cdot s)^{-1})$ in SI units and roentgen per second $(R \cdot s^{-1})$ in the former units. The unit of absorbed dose and of kerma can be applied to any material. Although all ionizing radiation can produce similar biologic effects, the absorbed dose necessary to produce a particular effect may vary from one kind of radiation to another and from one type of biologic material to another.

3.4 Relative Biological Effectiveness and Radiation Protection

The ratio of the absorbed dose of photons of a specified energy to the absorbed dose of any other ionizing radiation required to produce the same biologic effect is called relative biological effectiveness (RBE). For example, suppose that an absorbed dose of 10 cGy (or 10 rads), from a beam of "fast" neutrons with an average energy of 200 keV produces the same biologic effect as an absorbed dose of 200 cGy (or 200 rad), from a beam of photons with an average energy of 100 keV. Then the RBE for the neutron beam in this example is:

$$\text{RBE} = \frac{200 \text{ cGy}}{10 \text{ cGy}} = 20 \tag{3.4}$$

Apparently then, neutrons of this energy have a biologic effect which is twenty times greater than that of photons. The RBE of any particular kind of ionizing radiation is a function not only of the energy of the radiation, but also of the type and degree of biologic effect and the nature of the tissue or organ under consideration. Because of this specificity, relative biological effectiveness is usually reserved for precise radiobiologic work, and in 1991, ICRP, followed in 1993 by NCRP defined two new quantities designed to assist in the evaluation of radiation effects as they apply to people, particularly to radiation workers. The first of these quantities is known as the radiation weighting factor w_R. This is a dimensionless factor which is meant to describe the differences between types and energies of ionizing radiation as they affect biologic tissue, particularly within the range of absorbed doses of concern in radiation protection. For this reason the radiation weighting factor bears a close resemblance to the definition of quality factor (Q) which is defined as an average relative biological effectiveness factor based on the macroscopic effects of radiation on the human organism. The quality factor as used in setting radiation protection standards, has been replaced by the radiation weighting factor. It is worth noting that in radiation protection work, the quality factor is a weighting factor applied to the absorbed dose at a point, whereas the radiation weighting factor applies to the absorbed dose averaged over a tissue or organ. Table 3.1 gives a list of the more commonly encountered types of ionizing radiation and the associated radiation weighting factors.

Table 3.1: Radiation Weighting Factors for Some Ionizing Radiations[a]

Type and Energy of Ionizing Radiation		w_R
Photons, electrons, positrons and muon		1
Neutron, energy	< 10 keV	5
	> 10 keV to 100 keV	10
	> 100 keV to 2 MeV	20
	> 2 to 20 MeV	10
	> 20 MeV	5
Protons, other than recoil proton and energy > 2 MeV		2 [b]
alpha particles, fission fragments, nonrelativistic heavy nuclei		20

[a]Adapted from NCRP Report Number 116 (1993)   [b]ICRP uses 5

The other recently defined quantity, known as the tissue weighting factor w_T represents the proportionate detriment of tissue or organ T when the whole body is irradiated uniformly. The sum of tissue weighting factors has been normalized to unity over the whole body to allow a more uniform estimation as described in the next section, to be made using them. The values of the radiation weighting factor are dependent only on the type and energy of the radiation and are independent of the tissue or organ. The tissue weighting factors, on the other hand, are independent of the type and energy of the radiation and depend only on the tissue or organ being irradiated. Table 3.2 gives a list of tissue weighting factors, which were developed from a population consisting of equal numbers of both men and women, ranging in age from young to elderly.

ICRP and NCRP specifically name the following twelve organs and tissues: adrenals, brain, small intestine, upper large intestine, kidney, muscle, pancreas, spleen, thymus and uterus. Both ICRP and NCRP recommend that if one of the remainder organs or tissues not explicitly listed as one of the twelve in Table 3.2 receives an equivalent dose (as described below) in excess of the highest such dose to any of the twelve organs or tissues named, then a weighting factor of 0.025 should be applied to that organ or tissue and a weighting factor of 0.025 to all the others.

Table 3.2: Tissue Weighting Factors[a]

Tissue or Organ	w_T
Gonads	0.20
Bone Marrow (red)	0.12
Colon	0.12
Lung	0.12
Stomach	0.12
Bladder	0.05
Breast	0.05
Liver	0.05
Oesophagus	0.05
Thyroid	0.05
Skin	0.01
Bone Surface	0.01
Remainder	0.05

[a]Adapted from ICRP Publication 60 (1991)

3.5 Equivalent Dose and Effective Dose

In 1982, ICRP introduced, and in 1991 refined, the idea of stochastic effects, those which are random or statistical in nature and have no apparent threshold, and non-stochastic, i.e., deterministic effects, which have a threshold after which an effect is certain to occur. These definitions were adopted by NCRP in 1993. In attempting to quantify deterministic effects, it is useful to note that most of the organs and tissues of the human body can sustain the loss of a substantial number of cells with no observable effect. However, when the number of cell deaths is large enough, the tissue or organ suffers a pronounced functional deficit. One of the effects of absorbed dose from ionizing radiation is to cause cell death. Hence, at low values of absorbed dose, there will be no visible harm from cell death, but after a certain threshold, harm will be clearly manifested. Deterministic effects then, are those certain to occur after the absorption of a given amount of ionizing radiation. It is much more difficult to quantify the short and long term effects which result from the modification rather than death of cells. Reproduction of cells which have been modified, but are still viable, may or may not result after a prolonged and variable delay, sometimes known as the latency period, in a malignant condition which is usual called a cancer. Effects manifest in the person receiving the radiation are known as somatic effects. Absorption of ionizing radiation is naturally one of the ways in which cells can be modified. From epidemiological studies, it seems that the probability of malignancy occurring from radiation increases with each increment of dose. There is no evidence to suggest that there is a threshold for this kind of random or chance event, but it is believed that it is roughly proportional to the absorbed dose when this is considerably below the deterministic effect threshold. If the damage has occurred in a cell whose function it is to transmit genetic information to the progeny of the exposed person, hereditary or genetic effects may be manifested in succeeding generations.

The values selected for the radiation weighting factors given above are meant to be representative of the relative biologic effectiveness of that specified type and energy radiation in inducing stochastic effects at low absorbed doses. Previously, in radiation protection work, a quantity called "dose equivalent", which is a derived quantity defined as:

$$H = D \cdot Q \cdot N \tag{3.5}$$

was used to estimate the biologic effects of the absorbed dose. In this formulation, H is dose equivalent, D is absorbed dose, Q is quality factor, and N is other modifying factors. H is a radiation protection term and, since it is derived from the quality factor, it represents the weighted absorbed dose *at a point*. The derived SI unit for dose equivalent is the joule per kilogram which has the special name sievert (Sv).[2] Dose equivalent, therefore, has the same basic units as absorbed dose and kerma, vis., joule per kilogram ($J \cdot kg^{-1}$). The special derived name was changed to emphasize the change in meaning. In the former system of measurement, the special derived unit defined for dose equivalent was known as the rem, an acronym for roentgen equivalent man. In order to emphasize the difference in viewpoint between the quality factor (Q) and the radiation weighting factor (w_R) another quantity known as "equivalent dose" and given the symbol H_T, should now be used. This quantity H_T, represents the absorbed dose *averaged over a tissue or organ* (as opposed to a point) and is weighted for the type and energy of the radiation deposited. The equivalent dose in any tissue or organ is given by:

$$H_T = \sum_R w_R \cdot D_{T,R} \tag{3.6}$$

where $D_{T,R}$ is the absorbed dose due to radiation R which is averaged over tissue T. The special derived SI unit for equivalent dose is also the sievert (Sv) as described above. Since the probability of stochastic effects depends not only on the type and energy of the ionizing radiation, but also on the organ or tissue involved, the "effective dose" has been defined

---

2. The sievert (Sv) is named for a Swedish physicist, Rolf Maximilian Sievert (1896–1966), who was one of the pioneers in the clinical application of radiation physics and who advocated the study of radiation effects and the establishment of radiation protection measures. The ICRU, which is charged with defining the units used in radiation applications, has adopted the sievert, and it is used by the ICRP and the NCRP.

which is sum of the weighted equivalent doses in all the tissues and organs of the body. The effective dose is given by:

$$E = \sum_T w_T H_T = \sum_T w_T \sum_R w_R \cdot D_{T,R} \qquad (3.7)$$

where H_T is defined above and w_T is the tissue weighting factor defined in the previous section. The special derived SI unit for effective dose is again the sievert. Recall that the sum of the tissue weighting factors was normalized to unity. It is now clear that this was done to ensure that a uniform equivalent dose over the whole body would give an effective dose numerically equal to the sum of the equivalent dose for each organ.

Additionally, note that the values of the radiation and tissue weighting factors are derived from the current status of radiobiology and hence may change from time to time. The latest set of values as recommended by ICRP and NCRP must always be used. However, the definitions of equivalent dose (in a single organ or tissue) and of effective dose (in the whole body) are not meant to be confined to any particular values of w_R or w_T. Both equivalent and effective dose are intended to be used in radiation protection which includes the assessment of risks only in general terms and are intended to provide a basis for estimating the probability of effects to the individual only for an absorbed dose well below the threshold for deterministic effects. If an estimation of the likely consequences of an irradiation to a known population is to be made, it would be better to use absorbed dose, the relative biological effectiveness of the specific radiations, and probability coefficients relating to the population in question.

The last task is to deal with any other factors, such as absorbed dose rate, which might modify the actual stochastic effect(s). In the previous formulation, N was used to provide a generic value for all other unspecified weighting factors. In the 1991 formulation by ICRP, followed by NCRP in 1993, the use of N is discontinued. In order to cover all exposure conditions not specifically falling under the radiation or tissue weighting factors, different values of the coefficients relating equivalent dose and effective dose to the probability of stochastic effects should be used.

3.6 Other Dosimetric Quantities

When a radionuclide is deposited internally, the absorbed radiation dose to organs or tissues will depend on how long the radionuclide remains. There are two ways in which unsealed radionuclides are eliminated from the human body, by radioactive decay and by physiological elimination. Hence the absorbed dose from an internally deposited radionuclide has a temporal component that must be considered. Both ICRP and NCRP recommend the use of the committed equivalent dose $H_T(\tau)$ as the time integral of equivalent dose rate in a specific tissue (T) after the deposition of the radioactive material has occurred. For a single radionuclidic deposition at time t_0:

$$H_T\tau = \int_{t_0}^{t_0+\tau} \dot{H}_T t \, dt \tag{3.8}$$

where $\dot{H}_T$ is the equivalent dose rate in the organ or tissue T at time t and τ is the integration period to be taken as 50 y (fifty years) after intake for a radiation worker and 70 y for a child. The committed effective dose $E(\tau)$ given by:

$$E\langle\tau\rangle = \sum_T w_T H_T\langle\tau\rangle \tag{3.9}$$

is just the sum for each internally deposited radionuclide of the appropriate tissue weighting factor multiplied by the committed equivalent dose. When τ is taken to be 50 y the above equation is written for a radiation worker as:

$$E(50) = \sum_{T=i}^{T=j} H_T(50)w_T + w_{T=\text{remainder}} \frac{\sum_{T=k}^{T=l} m_T H_T(50)}{\sum_{T=k}^{T=l} m_T} \tag{3.10}$$

where $H_T(50)$ is the committed equivalent dose and w_T the specific weighting factor for the twelve organs or tissues (T_i to T_j) specifically named in Table 3.2 and m_T is the mass of the named remainder organs (T_k to T_l). If half of the original radionuclide is eliminated from the body in three months or less, then the committed equivalent and effective dose reflect very closely the annual equivalent and effective dose for the year of intake. However, if the radionuclide leaves the body more slowly, these quantities actually overestimate the equivalent and effective dose received in the year of intake because a significant portion of the absorbed dose is deposited in succeeding years. Thus these quantities are not recommended for estimating health effects or assessing the probability of deleterious effects. Although, they are useful for routine radiation protection purposes such as assessing compliance with the annual effective dose limits (to be given in the next chapter) and also in the design of facilities. The unit of committed equivalent and effective dose remains the sievert. The above quantities all relate to estimating the biologic effects of absorbed dose in the individual.

ICRP has defined additional quantities, collective equivalent and effective dose to estimate biologic effects in exposed groups or populations. This is achieved by multiplying the average absorbed dose to the exposed group by the number of persons in the group. If more than one group is involved, the total collective quantity is the sum of the collective quantities for each group. The unit of these collective quantities is termed the "man sievert". These collective quantities should be thought of as representing the total stochastic consequences to the group, but they should only be applied when the consequences are truly proportional to both the dosimetric quantity and to the number of people exposed. ICRP has defined other quantities such as the "dose commitment" to be used only as a calculational tool, the ambient and direction dose equivalent to be used for environmental and area monitoring and the individual dose equivalent, penetrating and superficial, to be used for purposes of monitoring individuals. It has also defined probability coefficients for relating probability of stochastic effects to dosimetric quantities. All of these are defined to aid in applying the principles of radiation protection to both those who are occupationally exposed and to the general public. For further information regarding these units, the reader is referred to ICRP Publication 60 [ICRP, 1991b].

Additionally, two more quantities are in common use. These are: excess relative risk (ERR) defined as the difference in the rate at which a health risk occurs in a radiation exposed population to that in an unexposed population, measured per gray (or per sievert); and excess absolute risk (EAR) defined as the excess number of cases per person-year (PY) per gray. These terms are used to express risk to exposed groups of people such as the Japanese atomic bomb survivors, the persons exposed as a result of the Hanover nuclear facility in the state of Washington, USA, and the persons exposed as a result of the nuclear power plant accident in Chernobyl, Ukraine.

3.7 Review Questions

1. Exposure is defined for what type(s) of ionizing radiation?

2. What is meant by electron equilibrium? Is it possible to establish it for photons of all energies? If not, what is the approximate energy range?

3. What is the special apparatus called which is used to standardize exposure?

4. For what type(s) of ionizing radiation is the absorbed dose defined?

5. Briefly describe what is meant by the absorbed dose rate.

6. What is the definition of kerma? For what type(s) of radiation is kerma defined? What is meant by the "interaction" site? by the "absorbed dose" site?

7. How is air kerma related to exposure? to absorbed dose?

8. What is the definition of kerma rate? of exposure rate?

9. What is meant by RBE? by quality factor? How do they differ?

10. What is meant by radiation weighting factor? by tissue weighting factor? How do they differ?

11. What is meant by stochastic effects? by deterministic effects? How do they differ?

12. What is the definition of equivalent dose? of effective dose?

13. What is the basis for using equivalent dose and effective dose in radiation

protection work?

14. What is the basis for using committed equivalent dose and committed effective dose in radiation protection work?

15. What are some other quantities defined by ICRP and NCRP for use in radiation protection work?

16. What are the special derived units and the special name for exposure? for absorbed dose and absorbed dose rate? for kerma and kerma rate? for equivalent and effective dose? collective equivalent and effective dose?

3.8 Problems

1. A radiation worker receives an air kerma in tissue (an exposure) of $1 \cdot 10^{-3}$ cGy due to photons whose energy is 100 keV. What absorbed dose in air is equivalent to this? What absorbed dose in tissue is equivalent to this?

2. A radiation worker receives an absorbed dose of fast beta particles of $1 \cdot 10^{-6}$ Gy in a period of 2 s. What is the absorbed dose rate?

3. If a radiation worker is exposed to an air kerma rate of $1 \cdot 10^{-4}$ cGy·s^{-1} due to low-energy photons for a period of 5 s, what is the total air kerma to which he/she was exposed? (Assume the exposure was uniform.)

4. A radiation worker receives an external absorbed dose from thermal (0.025 eV) neutrons of $3 \cdot 10^{-5}$ Gy. What is the equivalent dose he/she has received? What is the effective dose to the skin?

5. A radiation worker receives an accidental absorbed dose of 0.05 cGy from neutrons 1 MeV in energy. What is the equivalent dose that he/she received? What is the effective dose to the whole body (assuming uniform exposure).

6. In a radiation field, the following circumstances apply:

    absorbed dose rate of photons 10^{-6} Gy·s^{-1}

    absorbed dose rate of alpha particles $4 \cdot 10^{-6}$ Gy·s^{-1}

    absorbed dose rate of 5 MeV neutrons $2 \cdot 10^{-6}$ Gy·s^{-1}

    If a radiation worker is exposed to this external field for 10 s, what equivalent dose would he/she receive? What would be the effective dose to the red bone marrow?

Chapter 4

BIOLOGICAL EFFECTS AND EFFECTIVE/EQUIVALENT DOSE LIMITS

While ionizing radiation has many uses, some of great benefit to mankind, it is also true that its beneficial use must be carefully weighted against possible deleterious effects. Over the years, much evidence has been accumulated to document that ionizing radiation is not only potentially dangerous, but in fact lethal, if misused. The lower limit on the radiation dose that will cause death in a very short time is the subject of a great deal of research. What concerns scientists, in particular, radiobiologists, even more is whether very low levels of radiation absorbed dose, extended over a relatively long time, cause adverse reactions in humans.

Several years ago, the United States National Research Council commissioned a study group to seek an answer to this question. The initial results of the study, published in 1972, are detailed in Biological Effects of Ionizing Radiation (BEIR) report. The BEIR Committee studied in particular the fate of the early X-ray workers, the atomic bomb victims of Hiroshima and Nagasaki, and other groups who were routinely or accidentally exposed to ionizing radiation. In their initial report, the BEIR Committee found a linear relationship between radiation effects and absorbed dose (see Figure 1.1). Curve A shows the initial (linear) report results. After further study, some members of the BEIR Committee concluded that the data showed no indication of any harmful effects from very low levels of absorbed doses of ionizing radiation from low LET sources such as photons. This trend, which some researches believe to be correct, was known as the "threshold" effect and is illustrated in curve B. Curve C shows the opposite effect; i.e., low levels of radiation, particularly radiation predominantly composed of neutrons, may be proportionately more harmful at lower than higher levels. Arguments on both sides of the question were presented in the BEIR III report (1980). The BEIR IV report (1988), dealt with the health risks of radon and other internally deposited alpha-emitting radionuclides. In 1986, a new BEIR committee was formed and was given the charge to conduct a comprehensive review of the biological effects of ionizing radiation focusing on information that had been reported since the conclusion of BEIR III. Additionally, this new committee was to provide new estimates, to the extend that available information permitted, of the risks of genetic and somatic effects in humans due to low-level exposure to ionizing radiation. Furthermore they were charged to consider the risk estimates from both internal and external sources of radiation and to document the procedure by which the risk estimates were derived. Their report is provided in BEIR V (1990) and it is largely from this study that ICRP and NCRP have derived their current set of recommendations for equivalent and effective dose limits. Finally, the BEIR VII (2006) report has been published. This report not only considers all the previous evidence, but additionally, the data available after the nuclear power plant accident at Chernobyl in 1986.

The United Nations Scientific Committee on the Effects of Atomic Radiation (UNSCEAR) has also made similar studies. Which of the three curves is actually correct is a question that is unlikely to be resolved in the near future, but a conservative combination of the linear (A) and threshold (B) is currently being used by International Commission on Radiological Protection [ICRP, 1991] and Unites States National Council on Radiation Protection and Measurements [NCRP,1993]. Since radiation workers can be routinely exposed to very low levels of radiation, this question is crucial for them. The quantitative question of what is an acceptable (safe) level of radiation exposure and, consequently, radiation absorbed dose has been attempted by a number of different commissions and committees. While there is no conclusive evidence for any one argument, ICRP and NCRP have examined the evidence and made studies of their own resulting in a refinement of the limits on effective and equivalent dose to reflect the very latest radiobiological data.

One of the major radiological exposure events of the last century was the accident at the nuclear power plant at Chernobyl in northern Ukraine. This accident released a variety of radionuclides into the atmosphere which resulted in dispersion over a very wide area. The major radionuclides released were I-131 (half-life 8 days), Cs-134 (half-life 2 years), and Cs-137 (half-life 30 years). The release of other radionuclides such as plutonium isotopes, Sr-90, and Am-241 were not significant.

There were approximately 600 workers present on the site when the accident occurred, of whom 134 received high absorbed doses (0.7-13.4 Gy). This group along with the emergency workers who responded within the first few days, suffered from radiation sickness, and 28 died in the first three months after exposure, with two more dying soon afterward. During the next two years, about 200,000 recovery operation workers received absorbed doses of between 0.01 and 0.5 Gy. This group has been closely followed looking for potential heath risks.

Outside of the immediately exposed areas in Belarus, the Russian Federation, and Ukraine, other countries in Europe and North America were affected by the accident. Absorbed doses in these places were at most equivalent to an effective dose of 1 mSv in the first year after the accident and progressively decreasing in subsequent years. The estimated absorbed dose over the lifetime of this population was estimated to be 2-5 times the first-year absorbed dose. As these quantities are comparable to an annual dose from natural background radiation, they are of little significance radiologically.

The UNSCEAR 2000 reports that fourteen years after the accident there were about 1800 thyroid cancers reported in those who were exposed at the ages of 0-18 years. There has been no increase in non chronic lymphoid leukemia (CLL) and no increase in solid tumors. The incubation period for non-CLL is two to twelve years and for solid tumors, 10 to 15 years. CLL is not thought to be caused by radiation. UNSCEAR found no unambiguous evidence of genetic effects among the large number of persons exposed throughout Europe.

The BEIR VII (2006) reports confirms these findings. It estimates that the incidence of thyroid cancer may rise to high as 5000, but so far there have been very few deaths attributed to this disease. The risk of thyroid cancer, especially in those five years or older at the time of the accident is decreasing. BEIR VII estimates a two fold increase in non CLL in workers who received an average absorbed dose of 150 mGy. Although this has not been seen, it would be hard to distinguish from natural occurrences of this disease. The risk of non CLL is expected to decrease as the incubation period has passed. Since the incubation period for solid tumors is ten to 15 years or more, some solid tumors might yet be a result of the specific exposure from Chernobyl, but if so, BEIR VII estimates that the result is unlikely to have enough statistical power in the exposed population to distinguish it from other natural occurrences. There is also no evidence of genetic effects or infertility in the general population.

4.1 Regulatory Agencies

The standards for maximum exposure have been suggested by several different agencies. One, already mentioned in reference to the BEIR report, is the Unites States National Research Council; others are the ICRP, the NCRP, and the International Commission on Radiation Units and Measurements (ICRU). These agencies have no law-making or enforcing power; they simply do studies and make recommendations. However, based on their recommendations, limits can be set by the appropriate law-making body. In the United States, several agencies have been charged with enforcing the standards that are set. On a national level, Congress has designated the Nuclear Regulatory Commission (NRC), formerly the Atomic Energy Commission (AEC), to regulate safety in all aspects that pertain to reactor products and more recently, accelerator products as well. Thus, the NRC would be concerned with safety in transport, possession, and use as well as worker safety. In some cases the NRC has entered into an agreement with a state, known as an "agreement state," in which NRC allows the state to enforce regulations within the state. In these cases, the state regulations must be at least as stringent as those of the NRC. Likewise a state has sometimes entered into an agreement with one of its cities, e.g., the State of New York and New York City.

For other sources of ionizing radiation, the individual states enforce the regulations through their health codes. In some cases, as in New York City, a city enforces the regulations under its own health code. Radiation workers are also governed by the Occupational Safety and Health Act (OSHA). The intent of OSHA is to protect the safety and health of all workers; radiation workers are no exception. Based on the knowledge gained from the study of radiation effects, the stated objectives of any radiation protection program should be to prevent deterministic effects by maintaining effective/equivalent dose limits below the estimated threshold limit and to limit the risk of stochastic effects, cancer and genetic effects, to as low a level as is consistent with societal needs, values, benefits gained and economic factors. These objectives can be met by adhering to the "As Low As Reasonably Achievable (ALARA)" principle (see section 4.5) set forth by ICRP and by applying effective/equivalent dose limits for controlling exposures.

4.2 Occupational Exposure

In considering the standards for individuals who are occupationally exposed, ICRP and NCRP believe that the current recommendations provide far lower occupational risk than is found in many occupations normally considered as "safe" industries. An occupationally exposed individual normally performs his work in a controlled area (i.e., an area to which access is posted as restricted) or has duties that involve exposure to radiation. Such a person has had formal training in radiation protection and is subject to appropriate controls. Individuals such as messengers, maintenance workers, nurses, etc., who have reason to occasionally enter controlled areas, are described by the NCRP as "occasionally exposed individuals." Provided that the controlling organization assures that the intended exposure of these individuals is the same as that of the general public, occasionally exposed individuals need not be knowledgeable concerning radiation protection practices.

An occupationally exposed individual then, also known as a "radiation worker" is one who chooses to work in radiation with the tacit understanding that under normal working conditions his/her health is not likely to be impaired, although there may be certain associated risks. This individual derives a tangible benefit from radiation in the form of his/her normal livelihood. This is considered fair or positively advantageous to the individual when the basic radiation control criteria as promulgated in qualified recommendations, such as those of ICRP and NCRP, are conscientiously maintained by the employing organization, and the individual concerned is made cognizant of the nature of the risks and acts in accordance with these regulations. Specific recommendations along with some examples of good working habits are given explicitly in Chapter 8.

4.3 Biological Factors

The factors affecting the response of man's biological system to radiation exposure are many and varied. The first factor affecting radiation response is age. Embryonic stages are known to be highly sensitive. In addition, there may be critical periods of increased sensitivity associated with certain organ development between birth and maturity. Radiation exposure can produce genetic effects, i.e., effects seen in the descendants of the radiation workers, only if the exposure occurs before termination of the reproductive period. Radiation exposure can produce somatic effects such as leukemia and cancers only if the remaining life expectancy of the individual is less than the latent period. Furthermore, it has been well demonstrated that if radiation is administered in small increments distributed over a longer time interval (referred to as fractionation), total absorbed equivalent dose much higher than those required to kill an organism in a single exposure can be tolerated. This is attributed to the ability of cells that have received sublethal damage to recover, in some cases completely, from the radiation effects. This recovery and its extent is very dependent upon cell type, the tissue being studied, the stage of mitosis or cell division (if any) when the cell is irradiated, the type of radiation used (cells are able to recover better from exposure to low-LET radiation such as photons), and the sensitivity of the organism. For purposes of radiation protection, ICRP and NCRP consider it safe to assume that in most cases, the effects of exposure to low-LET radiation will be reduced by absorbed dose fractionation.

Exposure to high-LET radiation, such as neutrons, has quite the opposite effect. In addition to the fact that a smaller absorbed dose produces the same absorbed equivalent dose and, hence, the same degree of effect, cells demonstrate little or no ability to repair or recover from sublethal damage caused by high-LET radiation and, hence, fractionation of the total absorbed dose has little or no effect. In the case of exposure to low-LET radiation, it has been experimentally determined that cells which are well-oxygenated are more radiosensitive. This effect is minimal in the case of high-LET radiation exposure. Knowledge of the manifestation of radiation effects in man depend to a large extent on the observation of accident cases and the effects of restricted areas or volume irradiation as in radiation oncology.

It is usual to separate radiation exposure effects into the categories of genetic and somatic effects. NCRP has defined the Genetically Significant Dose (GSD) as the dose equivalent which, if received by every member of the population, would be expected to produce the same total genetic injury to the population as does the actual equivalent dose received by various individuals. From the kind and number of diagnostic examinations established by the Unites States Public Health Service, NCRP has estimated the GSD from medical irradiation to be between 500 to 700 µSv per person per year. UNSCEAR reports that the GSD in Great Britain of 118 µSv in 1977 was not significantly changed from the 141 µSv reported in 1957. Although it has never been demonstrated to cause hereditary effects in human populations, even in children of survivors of the atomic bombings in Japan, experimental studies have shown that radiation can induce hereditary effects in plants and animals However, UNSCEAR (2001) determined that better methods now exist for estimating the hereditary risks and it has concluded that:

> "for a population exposed to radiation in one generation only, the risks to the progeny of the first postradiation generation are estimated to be 3,000 to 4,700 cases per gray per one million progeny; this constitutes 0.4 to 0.6 per cent of the baseline frequency of those disorders in the human population."

As can be seen from the above, a significant degree of uncertainty exists in the estimation of the genetic consequences of radiation exposure in man. Hence, radiation exposure guides for persons of reproductive age are conservative. Individuals, whether radiation workers or the general population, receive exposure to photons only under conditions that result in low mutation rates. The risk is reduced if the absorbed dose is kept low; or if the absorbed dose rate is high, the total absorbed dose at a single exposure is kept small. Should a large single exposure be sustained as in a radiation accident, if conception is deferred, it is likely that most genetic effects can be avoided. The exact period of deferment is not stipulated for women, but is about two months for men. Injury to the reproductive organs are of natural concern to radiation workers. Permanent histological sterility, the complete absence of gametes, requires absorbed doses of photons to the gonads of 3.5 to 6 Gy for men and 2.5 to 6 Gy for women. Men may be sterilized permanently by localized irradiation without prominent changes in interstitial sex cells, hormone balance, libido, or physical capability. Women sterilized by radiation undergo an artificial menopause similar to natural menopause, but in young women, ovulation and normal hormonal balance is restored about one year after a whole body absorbed dose that is low enough to permit survival. Temporary sterility occurs with lower absorbed doses. A single absorbed dose of 0.15 Gy (low-LET radiation—delivered at a high dose rate) reduces sperm count in men by about 30%, starting six weeks after exposure with recovery occurring in about forty weeks. A single absorbed dose of about 0.50 Gy to the gonads may induce brief, temporary sterility in many men and some women; a single absorbed dose of approximately 2.5 Gy may induce temporary sterility for one or two years; and 5 to 6 Gy may induce permanent sterility, especially in persons with borderline original fertility. Note that these absorbed doses are much higher than any occupationally exposed person would receive in the normal course of his/her work.

Somatic effects are those which may become evident in the irradiated individual. Of these, neoplastic diseases in the form of leukemia and tumors (not all of which are malignant) are perhaps the most common. It is also well-known that sufficient exposure of the lens of the eye to ionizing radiation such as photons, beta particles, and neutrons may cause cataracts. The exact amount of exposure in each case is unknown, but it is believed that a "practical threshold" exists for induction of vision impairment in man by low-LET radiation, and that this is not so for high-LET radiation such as neutrons. From a limited number of studies, NCRP concluded in 1993 that microscopic distribution of energy deposition is more descriptive of risk than is average energy deposition. In addition to leukemia and solid tumors, other, non-cancer somatic effects have been attributed to overexposure to ionizing radiation. These include heart disease, stroke, and diseases of the digestive, respiratory, and hematopoietic systems. While there is strong epistemological evidence suggesting this, it is has not been conclusively proven.

It has also been shown that when radiation levels much higher than those stipulated by ICRP and NCRP for occupational exposure are encountered, there are serious effects on growth and development. This has been the rule in Japanese and other children studied. According to both BEIR V and the 1993 NCRP study, there was no increase in cancer due to in-utero irradiation, but mental retardation does result. The highest period of sensitivity is from the eighth to the fifteenth week where IQ is diminished by as much as thirty points per sievert. The deficit is less if the irradiation occurs between the sixteenth to the twenty-fifth week and is essentially zero before the eighth and after the twenty-fifth week.

At this time there is no firm basis for establishing a quantitative relationship between life shortening and absorbed dose in man. Existing knowledge from past and present experience with radiologists and with those employed in nuclear energy programs indicated that life shortening of persons whose exposure is maintained within presently recommended occupational limits would be too small to detect in the presence of so many other variables. Although current epidemiological studies have shown no measurable quantitative results, NCRP believes that it is probable that some small degree of life shortening in man may occur following radiation exposure resulting in a high absorbed dose. The Life Span Study (LSS) being conducted on the Japanese bomb survivors and now on the population exposed due to the accident at Chernobyl, may be able to provide more evidence for or against life shortening.

4.4 Manifestation of Overexposure in Adults

Biological manifestations of overexposure to ionizing radiation are divided into early (acute) effects, secondary or delayed effects, and late effects. Early radiation effects, such as white blood cell changes, vomiting, and bone marrow depression occur only after absorbed doses at least as high as 0.25 Gy for the whole body measured near the midline. The midline whole body absorbed dose sufficient to cause death to 50% of the cases within 30 days (written as LD_{50}^{30} in which LD signifies lethal absorbed dose) is between 2.60 and 3.25 Gy of photons (this corresponds to a whole body absorbed dose of 4 to 5 Gy). If the person survives the acute effects of radiation exposure, secondary effects such as skin ulceration or suppression of thyroid hormone secretion, which results from destruction of normal bronchial and vascular tissue and subsequent attempts at repair, are observed. If a person who suffers substantial overexposure recovers from the early or acute effects and accommodates to the secondary or delayed effects, he may still be subject to late effects especially in the form of neoplastic diseases (primarily leukemia and tumors).

A distinction should be drawn here between a single absorbed dose of whole body radiation of 2 Gy received at a high absorbed dose rate (i.e., within a few minutes or hours) and an accumulated absorbed dose of the same magnitude distributed over an occupational lifetime. The former would be expected to put the person so irradiated at some incremental risk for neoplastic disease, especially leukemia, while the latter would not. If only part of the body is accidentally overexposed, the results are expected to be essentially the same as those given above if an entire organ system is involved. If only part of an organ system is involved, the results may or may not be mitigated depending on the circumstances.

The brief overview of the biological effects of radiation given here is meant simply as an introduction to this subject for the radiation worker. For more details, the reader should consult the references provided in the bibliography and other suitable works.

4.5 Recommendations for Effective/Equivalent Dose Limitations

In ICRP Publication 9 (since superseded by 22, 26 and 60), the ICRP strongly recommended that the limits set forth by itself and others be regarded as acceptable limits. It does, however, urge that "any unnecessary exposure be avoided and that all absorbed doses be kept As Low As is Readily Achievable (ALARA), economic and social considerations being taken into account." The ALARA statement has been widely used in radiation protection programs. In Publication 22, ICRP changed "readily" in the ALARA statement to "reasonably", and reaffirmed the basic recommendations of Publication 9. In Publication 26 and reiterated in Publication 60, the ICRP urged that the ALARA principle be regarded as the acceptable limit for radiation protection standards. This publication and several succeeding publications of the ICRP reaffirm and give practical examples for the implementation of this principle. The ALARA principle has been endorsed by the NCRP and adopted in the United States by the NRC as the basic limit by which individual radiation exposure and facility design will be judged. Limits of exposure have been set for occupationally exposed persons and members of the general public. The amount of exposure given to an individual in the course of diagnostic or therapeutic procedures is regulated mainly through accepted medical practice, and is not included in an individual's effective/equivalent dose limit because the benefit derived is believed to outweigh the physical risks associated with it. Natural background exposure is also excluded from this recommendation. However, the special case of radon levels in homes has been addressed by both the ICRP and NCRP. There is no simple solution to this problem in terms of established dwellings and societal cost. Newly built dwellings can be required to address radon hazards. NCRP recommends remedial action levels for the public to maintain a 5 mSv annual average effective dose for exposure from natural sources excluding radon and an annual average of $7 \cdot 10^{-3}$ J·h·m^{-3} for total exposure to radon and its decay products.

The dose limits are stated in terms of effective or equivalent dose (whichever properly applies to the particular limit) and are intended to be the sum of the effective/equivalent dose from external radiation and the committed effective/equivalent dose from internal deposition of radionuclides. There is no longer a population limit set by any recommending or governing body. The genetic risk to an individual is now included in the recommendations regarding stochastic effects. Because this genetic consideration is very important, it has been studied extensively. However, it is usually the province of radiobiology to assess risks and benefits to the population as a whole from ionizing radiation. Remember that a certain number of mutants are born in the normal course of events. It is possible though, exposure to ionizing radiation may increase the frequency with which such mutants are born, but this has not been epistemologically confirmed.

While the effective/equivalent dose limits as recommended by ICRP and NCRP are stated by these agencies and reproduced here in terms of "mSv", the reader may convert to "rem" by simply dividing the number by ten (1 mSv = 0.1 rem).

In deriving an effective dose limit for occupational exposure, both ICRP and NCRP considered that a single uniform whole body equivalent dose of 0.1 Sv would result in an average nominal lifetime excess risk of $4 \cdot 10^{-3}$ for fatal cancer, of $0.8 \cdot 10^{-3}$ for nonfatal cancer and for severe genetic effects, making a total risk of $5.6 \cdot 10^{-3}$. Also considered were the facts that radiation induced cancers occur late in life while accidental deaths in the workplace can occur at any time. Therefore, it was judged that for those few individuals who might receive effective/equivalent doses close to the limit over their working life, their total lifetime detriment incurred each year should be no greater than the annual risk of accidental death for a worker at the top end of the safe worker range. The annual effective dose limit based on stochastic effects for an occupationally exposed individual is 50 mSv. ICRP adds to this limit a further cumulative effective dose limit of 100 mSv in 5 years (y) (average annual value is therefore 20 mSv), while NCRP sets the cumulative effective dose limit as 10 mSv × age (y). The ICRP regulations, therefore, imply that an individual's radiation records need only be kept for five years, while those of NCRP imply that these records be kept for life.

The annual equivalent dose limit based on deterministic effects for an occupationally exposed individual is 150 mSv for the lens of the eye and 500 mSv for the skin, hands and feet. The annual effective dose limit based on stochastic effects for a member of the general public is 1 mSv for continuous exposure. ICRP adds that higher values may be incurred for a given period provided that the annual average over 5 years does not exceed 1 mSv, while NCRP sets a limit of 5 mSv for infrequent exposure. The annual equivalent dose limit based on deterministic effects for a member of the general public is set by ICRP and NCRP to be 15 mSv for the lens of the eye and 50 mSv to the skin, hands, and feet. ICRP sets a 2 mSv equivalent dose limit to the woman's abdomen once pregnancy has been declared and limits the intake of radionuclides to 1 mSv. The NCRP recommends a monthly equivalent dose limit of 0.5 mSv to the embryo-fetus once pregnancy is known. Note that the Unites States NRC has stated that a radiation worker must "declare" her pregnancy before she can be treated as a "pregnant radiation worker". This is in concurrence with the ICRP recommendations. NCRP has defined an effective dose of 0.01 mSv as a Negligible Individual Dose (NID). This establishes a lower limit below which the effective dose from any source or practice can be dismissed from consideration. For trainees under 18 years of age, an annual effective dose of less than 1 mSv, an equivalent dose to the lens of the eye of less than 15 mSv and to the skin, hands, and feet of less than 50 mSv is recommended. The annual effective dose of less than 1 mSv is considered to be part of the annual limit of 5 mSv for infrequent exposure to the members of the general public and not supplemental to it.

For occupational workers, ICRP recommends Annual Limit of Intake (ALI) and NCRP recommends Annual Reference Levels of Intake (ARLI) which limit the committed effective dose from a radionuclidic intake in a single year to the annual average (ICRP) value of 20 mSv named above. The ALI (ARLI) for any radionuclide is obtained by dividing the annual average effective dose limit (20 mSv) by the committed effective dose, E(50), resulting from an intake of 1 Bq of that radionuclide. Occasional intakes slightly greater than 2.5 times the ALI or ARLI are considered to have little effect on the long-term health of the individual involved. However, such occurrences usually indicate that improvements in the facilities or operating procedures are warranted.

Additionally, NCRP recommends Derived Reference Air Concentration (DRAC) as that concentration of a radionuclide which if breathed by Reference Man (see Chapter 11) inspiring 0.02 m^3 per minute for a working year, would result in an intake of one ARLI. Thus:

$$DRAC = \frac{ARLI}{40 \text{ h} \cdot \text{week}^{-1} \cdot 50 \text{weeks y}^{-1} \cdot 60 \text{min} \cdot \text{h}^{-1} \cdot 0.20 \text{ m}^3 \cdot \text{min}^{-1}} \tag{4.1}$$

DRAC's purpose then, is to provide a method for controlling exposures in the workplace to the ARLI. Values for DRAC apply to individual radionuclides and hence must be reduced appropriately if more than one radionuclide is involved. While DRAC does not apply directly to other means of radionuclide intake such as water, or to members of the general public, it can be used to derive limits for these situations as well when the difference in effective/equivalent dose limits and other variables are taken into consideration.

Further consideration must also be given to exposures in excess of the current recommendations which may occur because of the change in the regulations themselves, may be tied to a hazardous industry, or may be incurred because of an emergency situation. In all cases it must be borne in mind that the recommendation of ICRP and NCRP are meant to reduce the lifetime risk incurred by an individual worker. Exceeding the annual effective dose limit in a given year can usually be readily offset by the past and future which result in a lifetime cumulative effective dose of 10 mSv × age (y). However, if it is not possible to offset, under the current recommendations, a past history of overexposure (for example, a worker has a cumulative effective dose well beyond the current limit of 10 mSv × age (y)) then a limit of up to an average of 100 mSv in 5 y or 50 mSv in any year may be agreed upon by both the employer and employee. Because the current recommendations were developed with "safe" industries as a guide, it may happen that there are some industries in which these limitations cannot be met, but in which the hazards of the workplace apart from radiation also exceed the standards for a safe industry. In these circumstances, special effective/equivalent dose limits could be established.

In an emergency situation, only actions involving life saving justify exposures which significantly exceed the annual effective dose limit. While only those persons who volunteer should be used, older workers with low lifetime accumulated effective dose should preferably be chosen from among this group. If such persons are to be exposed to an equivalent dose of 0.5 Sv or more, they should not only be appraised of the acute effects, but also of the increase in their lifetime risk of cancer. When the emergency action is not life saving, it should be controlled to annual effective dose limits if possible. If this is not possible, both ICRP and NCRP recommend that an effective dose of 0.5 Sv be applied as well as an equivalent dose of 5 Sv to the skin. Table 4.1 summaries the recommendations of ICRP and NCRP. It should be noted once again that these recommendations exclude irradiation by both natural sources and medical procedures.

Table 4.1: Recommendations for Annual Effective/Equivalent Dose Limits[a]

	ICRP	NCRP
A. Occupational Limits:		
1. Effective dose limits		
a. Annual	50 mSv[b]	50 mSv[b]
b. Cumulative	100 mSv in 5 y	10 mSv·age (y)
2. Equivalent Dose Annual Limits		
a. lens of the eye	150 mSv	150 mSv
b. Skin, hands and feet	500 mSv	500 mSv
B. Public Limits (Annual):		
1. Effective dose limit		
a. continuous or frequent irradiation	1 mSv	1 mSv
b. infrequent irradiation	5 y average < 1 mSv/y	5 mSv
2. Equivalent dose limits		
a. lens of the eye	15 mSv	15 mSv
b. Skin, hands and feet	50 mS	50 mS
4. Remedial action for natural sources:		
a. Effective dose (excluding radon)	—	>5 mSv
b. Radon Irradiation (decay products)	—	$>7 \cdot 10^{-3}$ Jhm^{-3}
C. Embryo-Fetus Limits:		
1. Equivalent dose limit	2 mSv (abdomen; after pregnancy declaration); 1/20 of an ALI limit on radionuclide intake	0.5 mSv monthly (after pregnancy known)
D. Educational and Training Limits(Annual):		
1. Effective dose limit	—	1 mSv
2. Equivalent dose limit		
a. lens of the eye		15 mSv
b. Skin, hands and feet	—	50 mSv
E. Negligible Individual Dose:		
1. Annual effective dose (per source or practice)	—	0.01 mSv

[a] Adapted from NCRP Report Number 116
[b] To find the recommended limit in rem, simply divide by ten. Recall that 1 mSv = 0.1 rem.

4.6 Use of Effective/Equivalent Dose Limits

The major use of the effective/equivalent dose limits is to derive working standards. One such standard applies to radiation facilities themselves and regulates the amount of radiation that may be present in areas where occupationally exposed personnel work and where nonoccupationally exposed people are present or work. (An example of the latter is a secretary's office or a waiting area.) Using the recommended effective/equivalent dose limits, a radiation worker should not be exposed to more than 25 μSv per hour or to more than 4 mSv per month. Therefore, ambient radiation levels from X-ray machines or from radionuclides must be kept below this level. The effective/equivalent dose limits are used to derive the maximum permissible body burden (MPBB) from chronic ingestion. Consider the radium-dial painters: many, actually most of them, contracted leukemia, lung, and/or bone cancer because they used to put a nice point on their paintbrush by placing it in their mouths. No longer are these materials hand painted onto the dials and hands of watches and clocks which are now made luminescent (i.e., made to glow in the dark) by the use of materials such as tritium which emits a beta particle having a maximum energy of 0.0186 MeV.

In addition, the effective/equivalent dose limits allow the determination of maximum permissible concentrations (MPC) of radiation in air and water and other aspects of the environment. These limits are taken into consideration when facilities are in the design stage, and shielding must be planned for use in and with these facilities in order to reduce exposure and consequently, absorbed dose to an acceptable level. Chapter 7 details the practical methods used to reduce exposure to the acceptable level. This quantity is also important with respect to the disposal of radioactive liquids and gases. Chapter 9 will discuss the disposal of radiative waste material.

4.7 Review Questions

1. What is the role of agencies such as the National Academy of Sciences, the ICRP, the United Nations, and the NCRP in setting standards for maximum exposure to ionizing radiation?

2. What agency, in the United States, carries the power, invested in it by Congress, to set radiation exposure limits? What kind of ionizing radiation can it control? Who controls the remainder?

3. What agency, in your country, carries the power to set radiation exposure limits? How did it obtain that power? What kind of ionizing radiation can it control? Who controls the remainder?

4. What is the definition of occupational exposure? Are all persons who have access to controlled areas necessarily occupationally exposed workers?

5. With respect to biological damage inflicted by ionizing radiation, what is the definition of genetic effects? Of somatic effects?

6. What type of radiation, high- or low-LET, is considered to inflict more biological damage?

7. What is the definition of GSD?

8. Into what three stages are the biological manifestations of over exposure to radiation divided?

9. What is the significance of the symbol LD_{50}^{30} ?

10. What is the significance of the annual effective/equivalent dose limits? What agencies have made these recommendations? Is medical exposure included? What about background exposure?

11. For what type(s) of irradiation is the committed effective/equivalent dose defined? What is its use?

12. What are some of the uses of the annual effective/equivalent dose limits?

4.8 Problems

1. A radiation worker has accumulated a lifetime effective dose of 50 mSv at age 28. What annual effective dose limit is this individual recommended to receive during the next calendar year?

2. A radiation worker has accumulated a effective dose of 2 mSv per year for each of the last ten years. If this person is now 38 years old, what is the maximum allowable lifetime effective dose if:

    (a) he/she had received no previous effective dose?

    (b) he/she has received an effective dose as stated above?

3. An occupational worker is irradiated over the whole body by a uniformly distributed absorbed dose rate from photons of 10^{-6} Gy·s^{-1}. If this occupational worker is irradiated for one hour (1 h = 3600 s) what is the effective dose received? Is this more than the yearly limit? Suppose the irradiation had been by alpha particles, would the annual effective dose limit have been exceeded?

Chapter 5

COMMON SURVEY AND
CALIBRATION INSTRUMENTS

Survey and calibration instruments, are an important part of any good radiation safety program. Such instruments must be chosen correctly for the particular application and used properly. In order to select the most appropriate survey or calibration instrument, it is necessary to understand the features that compose a good working instrument. For area, container, or population survey monitoring, it is important to know the energy dependence and the sensitivity of the various types of monitors available. The early detection of a radiation hazard can be instrumental in the prevention of unnecessary exposure (air kerma) accumulation by radiation workers and others.

In this chapter we review the use of the radiation detectors studied in Chapter 2 for surveying physical locations, containers and emergency sites, as well as for calibrating diagnostic and oncologic radiation generators. In the next chapter we shall discuss the use of these instruments for monitoring the exposure (air kerma) of personnel.

5.1 Survey Monitors

One of the most important ways in which the detectors described in Chapter 2 are used is as survey monitors, survey instruments, or laboratory monitors. The purpose of a laboratory monitor or survey instrument is twofold: to keep constant surveillance over the working environment and to detect the quantity and extent of radiation contamination of that environment. Consequently, laboratory monitors are placed in rooms where radionuclides are or may be present. In addition, counters, floors, and other places where radionuclides are used are surveyed daily, while rooms containing generating equipment for diagnostic or oncologic quantities of radiation are surveyed periodically for leakage and scattered radiation. Areas where the radiation level may rise above the safe level due, for example, to leakage from a nearby radiation source (such as a nuclear reactor), malfunctioning of a collector-disposal system (such as a ventilation system), or incorrect output from a radiation source (such as an oncology or fluoroscopy generator, which may either remain on when switched off or generate a higher radiation level than indicated, or a waste container or delivery vehicle) are often provided with survey monitors containing alarms.

5.1.1 Characteristics of an Effective Survey Instrument

Survey instruments must possess the following characteristics: simplicity of construction, ruggedness, reliability, portability, and sensitivity. An instrument that is simply constructed is not only easy to use, but its circuitry is easy to comprehend, its parts are readily replaceable, and its components are conveniently arranged. These features also enable the survey instrument to be reasonably priced. Since it will most likely be handled by many different people, it should also be rugged enough to survive mild abuse.

Reliability is another important quality of a survey monitor. As will be discussed later in this chapter, survey monitors, particular Geiger counters (also known as Geiger Müller or G-M tubes), must be calibrated regularly. In order to ensure that the reading obtained is a correct one, a small source of radiation (known as a check source) is usually provided with this type of instrument. When the detector is placed near the check source, a precalibrated reading should be obtained. This procedure should be done immediately prior to use since one of the major functions of a survey monitor is to provide an accurate estimate of the amount of radiation present. Portability is desirable because it is often necessary to use a given survey instrument in a number of different locations. For this reason, the survey monitor should be light and compact and should contain its own battery.

Finally, the main reason why a survey instrument is present in the area is to monitor contamination or leakage of radiation; consequently, it must be very sensitive to both the type of radiation being monitored and its energy range. For example, in most instances it is necessary to detect photons and electrons with energies between 10 and 1000 keV, or alpha particles above 3 MeV in energy. Hence a Geiger-Müller tube detector, ionization chamber, or scintillation counter is usually used. G-M tubes and scintillation counters make excellent monitors because of their high sensitivity to photons and electrons in the energy range being considered. On the other hand, if neutrons or low-energy photons or beta particles are to be detected, a proportional counter should be used.

Generally speaking, no one survey instrument will have all of the characteristics described, so it is important to select the proper instrument for a particular monitoring situation. Further, it is necessary to know how to use that instrument intelligently and to be able to interpret the results.

5.1.2 Energy Characteristics

An important category of the survey instruments encountered in radiation protection work are gas-filled detectors. Instruments of this type operate by virtue of the ionization produced in the gas by the radiation to be detected. In general the instrument gives a reading in counts per minute (cpm), counts per second (cps), exposure (R), exposure rate (R/h), equivalent dose (mSv or rem), or equivalent dose rate (mSv/h or rem/h). All of these measurements are designed to be an indication of how much radiation is present. Gas-filled detectors are usually energy dependent; i.e., the measurement obtained depends not only upon the amount of radiation present but also the energy of that radiation. Instruments with good energy characteristics give readings based on the amount of radiation present only and are not influenced by the energy of that radiation.

Energy dependence occurs for two reasons. The first has do with the material used to make the walls of the detector. Let us consider that the detector is filled with air and that the amount of radiation present is determined relative to the ionization of air. Consequently, the walls of the chamber in which photons are attenuated should be made of an air-equivalent material, i.e., a material which has the same or nearly the same atomic number (Z) as air. (The Z for air is approximately 7.64.) At the same time, the thickness of the walls is a second factor to be considered. The probability that a photon will interact with the material composing the detectors walls and the penetrating ability of particles emitted, are functions of energy. (Recall from Chapter 1, the methods by which photons interact with matter.) Very high-energy photons have a smaller interaction probability (known technically as the interaction cross section), i.e., they tend to interact less than lower energy photons; hence, the walls of the chamber should be thick. On the other hand, very low-energy photons produce particles that, because of their small penetrating power, do not reach the gas-filling to be detected if the chamber walls are too thick. Thus the thickness of the chamber walls influences the number of photons detected. Ideally, a survey instrument with adjustable walls is best for learning about the energy composition of the radiation. We shall see later in our discussion of calibration instruments for radiation sources that precision electrometers, capable of accepting different ionization chamber probes for detecting the amount of radiation in different energy regions are just such devices.

5.1.3 Ionization Chamber Survey Monitors

Ionization chambers can be constructed so as to be useful for any kind of radiation monitoring. They are best used as photon measuring instruments, but they can be modified to monitor for alpha, beta, and even neutron radiations. Furthermore, although ion chambers have less sensitivity than Geiger counters, they can be used in high counting-rate situations, where they give precise results. Since ion chambers have good energy-dependence characteristics, they are desirable instruments for general radiation-safety, instrument calibration, and survey work.

Ionization chambers, when used for radiation survey work, always operate in mean-level mode. Energy absorbed from the radiation ionizes the gas in the chamber. Either the total charge produced or the total charge produced per unit time is measured. If the gas ionized is air and if the radiation consists of photons, it is conventional to speak of exposure. If the total charge produced is measured, the quantity found is exposure. If the charge per unit time is measured, the quantity found is exposure rate.

Ionization chambers used as area monitors usually employ unsealed air detectors to maintain the pressure at one atmosphere (101.3 kPa) for ordinary room temperature conditions. Some chambers are sealed, but in either case, the instrument must be allowed to come to equilibrium with the environment in which it is to be used. If subjected to a large temperature differential due to being transported between sites, a certain "warm-up" time must elapse before an accurate reading can be obtained. This warm-up time is specified for each instrument.

Survey monitor results are presented in two ways. On analog instruments the results are read by observing the deflection of a needle relative to a fixed scale. Other instruments present the results digitally, usually on a liquid crystal display (LCD); these instruments are generally capable of being interfaced (either through an RS232 or wireless connection) to a computer for data storage and analysis. Both kinds of instruments invariably respond in units of exposure rate or equivalent dose rate; the maximum rate measurable on the scale is usually switch selectable. In most instances, the ion chamber detector itself, is mounted external to, or on one side of, the measuring apparatus. Some, however, are internally mounted. These chambers can be sensitive to alpha particles with energies as low as 3.5 MeV, to beta particles with energies as low as 70 keV and to photons with energies as low as 6 or 7 keV.

To measure only higher energy photons (starting between 25 and 40 keV) exclusive of alpha and beta particles, most chambers can be fitted with a cap or with a sleeve. Some detectors are special purpose units shielded from radio frequency radiation (RF) and these only measure photons above 40 keV and beta particles above 200 keV in energy. There is one special purpose unit, shielded from RF, which measures only low-energy photons (between 12 and 50 keV) and betas above 150 keV. With these exceptions, most ion chamber survey meters can measure photons of energies up to 100 to 300 keV mixed with alpha and beta particles without cap or sleeve, and energies from 40 keV to 1.3 or 2 MeV exclusive of alpha and beta particles with a cap or sleeve in place. Most ion chambers used for survey do not include a check source, but when it is included, it is usually a depleted uranium source less than 370 kBq (10 µCi) in strength. These instruments must be checked periodically with a known, calibrated source to ensure proper function. The measuring apparatus is usually run by one or more flash light batteries (called "D" cells). In some instances (more often in G-M counters) the "D" cells are used to drive a solid state electronic circuit which maintains the detector tube voltage as well as powers the measuring apparatus. The battery shelf life tends to be about one year.

5.1.3.1 "Cutie Pie" Monitors

As a concrete example, let us discuss in some detail a popular ion chamber survey monitor known as the "cutie pie". The particular instruments, shown in Figure 5.1, have two overlapping ranges which can be switched between integral mode, where the results can be read in subdivisions of Sv or R, and count rate mode where the results are expressed in subdivisions of Sv/h or R/h. The precision of these instruments are rated at 10% of full scale indications over the total temperature, humidity, pressure, and operating range. Ninety percent of the final scale reading on the lowest range is equilibrated in about 2.5 s on the lowest scale and falls to 2.0 s for the higher range, which shows that response to higher exposure rates is faster. The instruments can detect alpha particles, beta particles, and photons. It is supplied with a cap to discriminate against alpha and beta particles, and low-energy photons. The instruments are essentially energy independent over their entire range.

The analog "cutie pie" can be used between 2 °C and +50 °C, and the digital version is essentially usable at all temperatures. After a 15 minute warm-up period, the instrument has a temperature dependence of less than 1.0% per °C at room temperature. The analog detector and the measuring apparatus are run by a Carbon-Zinc, mercury or nicad type battery. The digital version requires 11 NEDA CR-1220 having a self life of 7 years and 6 AA batteries, having a shelf life of 1000 hr. A battery check function is supplied.

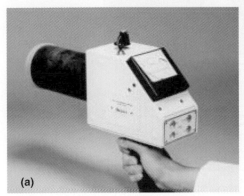

(a)   (b)

Figure 5.1: A typical analog "cutie pie" Exposure Rate Meter (a). A typical digital "cutie pie" Exposure Rate Meter (b). (Courtesy of Pinestar Technology, Inc.)

5.1.3.2 Digital Ionization Chamber Monitors

The instrument shown in Figure 5.2 is available in a non-pressurized air chamber version (shown) or as a pressured air chamber. The ion chamber survey meter features a slide for reading low energy beta particles and can read alpha particles above 4 MeV, beta particles above 100 keV and photons above 7 kev. It has five different overlapping operating range for a full scale value of 0-50 R/h or 0-500 mSv/h. Its accuracy is rated as 10% of the full scale indication at standard pressure and temperature (101.3 kPa, 22 °C). The instrument's energy dependence is shown in graphical format. This instrument can be operated with full specifications between -40 and 70 °C and is generally moisture resistant. The warm-up time is one minute. The instrument is powered by two 9 V alkaline batteries. The calibration source is ^{137}Cs, and there is an internal control for calibration.

5.1.4 Geiger-Müller Monitors

Geiger-Müller tubes are extremely useful for monitoring weak sources of beta and gamma radiation. Because the tubes can be made in virtually any size or shape, and because of their high sensitivity, Geiger counters are the instruments of choice for monitoring contamination and searching for lost radiation sources. Most G-M tubes are equipped with an audible sound system, which is especially desirable because the user does not need to

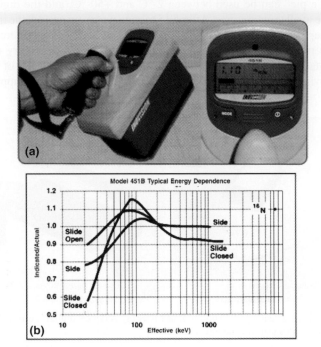

Figure 5.2: A typical digital ion chamber survey meter (a); the graph shows the energy dependence (b). (Courtesy of Fluke Biomedical.)

read the meter continuously in order to be aware of or to locate a radiation source.

Geiger-Müller tubes are used to monitor photons and alpha and beta particles. On the whole, these tubes are operated in pulse mode; consequently, the measuring apparatus takes the form of a count-rate meter with readings usually given in counts per minute (cpm). There are some measuring devices which read exposure rate (mR/h or R/h) or equivalent dose rate (mSv/h). The relatively long dead time of G-M tubes makes them unsuitable for accurate counting at high counting rates, but their high sensitivity makes them ideal for alerting the user to a, possibly unsuspected, source of irradiation. Geiger-Müller counters used as area monitors generally employ sealed tubes which are halogen quenched. There are some organic quenched tubes still available, but these should be avoided as noted in Chapter 2, Section 2.1.2.3. The tubes are generally external to the measuring apparatus and tubes with different characteristics are usually interchangeable (i.e., can be used with the same measuring apparatus — as shown in Figure 5.3). A few monitors, usually special purpose, such as for detecting low-energy photons, from ^{125}I for example, with activities below 100 Bq, have the detector chambers built inside. Some Geiger tubes used for beta as well as gamma detection, have sliding shields or end caps to discriminate against beta particles.

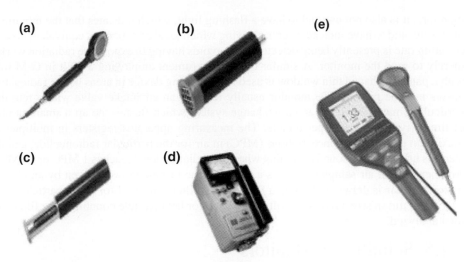

Figure 5.3: A typical pancake style for area and hand monitoring (a), an end window (b), and a side window (c), G-M tube. A typical analog survey meter (d) and a typical digital survey meter (e) that can be used with either a G-M tube or scintillation detector. The probes shown in (a-c) can be easily removed and interchanged with another in order to adapt the instrument to the particular radiation type and count rate to be surveyed. (Courtesy of Pinestar Technology, Inc. (a-c), Ludlum Measurements, Inc. (d), and Fluke Biomedical (e).)

In general, alpha particles above 3.5 MeV, beta particles above 35 keV (sometimes as high as 200 keV minimum), and photons with minimum energies between 6 and 12 keV can be detected with these systems. The survey meters come in various models which have various counting rates in counts per minute (cpm) and various exposure rates available in mR/h or equivalent dose rates in mSv/h. The response time of these instruments is fast; 90% of the final reading is obtained in no less than 15 s on the lowest scale. Some response times are as good as 4.5 s on the lowest scale. The measuring apparatus as well as the tube potential for these instruments is powered by two standard flashlight batteries ("D" cells). The conversion of the "D" cell voltage to operating voltage for the tube is accomplished through solid state circuitry. A check source (usually of depleted uranium) should always be provided. This is especially necessary for G-M tubes as they tend to easily lose their calibration. These monitors are generally operated under ordinary room temperatures, but they can be operated over as wide a temperature range as −20 to +50 °C with a temperature dependence within 0.2% per °C. One special feature of G-M survey monitors is the provision of an audible alert either through a loud speaker or earphone. For general survey work, most instruments have the audio signal follow the count rate fluctuation.

In some industrial situations, as well as in radiation oncology applications, it is often necessary to have not only a visual area monitor and warning alarm, but a remote alarm as well. G-M detectors are ideal for this situation because of their sensitivity and reliability. Special purpose instruments have an audio signal that can be set to respond only after a certain radiation level has been reached. This alarm trip level setting is usually internally mounted to prevent accidental or unauthorized changes. On such warning/alert monitors, it is also not unusual to have a flashing light, which indicates that the monitor is operating, and to have indicator lights showing which level of exposure, equivalent dose or counting rate is presently being detected. This avoids having to expose the radiation worker merely to read the monitor. A similar type of instrument employing a built in G-M tube with a pancake shaped thin window is used as an alerting device in areas where radioactive gasses may be present. The monitor usually features an air-intake valve with a reusable particulate-matter filter and an air exchange system which flushes the air a small number of times (typically three) per minute. The measuring apparatus registers in multiples of maximum permissible concentrations (MPC) in air for the particular radionuclide, and the alarm is usually set to sound a warning when the radiation level reaches 1 MPC or 10 MPC. Another type of air sampler employs a high volume turbine blower driven by an electric motor. The air is drawn in through a filter. Paper filters are used for alpha particles from 0.01 to 10 μm in size. Glass fiber filters are used for beta particle counting. The filters are then analyzed.

5.1.5 Scintillation Monitors

The ability of scintillation detectors (because of their short recovery time, typically 3 to 4 μs in this application) to accurately determine very high counting rates and to detect the presence of even small amounts of radioactivity makes them useful for monitoring areas where such levels might occur. Their energy discrimination ability is also desirable in some applications. Like G-M tubes, the scintillation detector can give an audible signal that can be used constantly or only as an alarm. Both the expense of the scintillation detectors themselves (the measuring apparatus is often interchangeable with those associated with the G-M tube, i.e., either probe can be affixed, see Figure 5.3) and their relative fragility militates against these instruments being as widely used as ionization chambers or G-M tubes.

Three types of scintillation probes are in common use for survey purposes. The first is a 2.54 cm diameter by 2.54 cm thick (1 inch by 1 inch) NaI(TI) crystal for measuring photons with energies above 90 keV and exposure rates up to 35 R/h. A thinner NaI(TI) crystal (2.54 cm in diameter by 1 cm thick) is used for measuring photons with energies between 10 and 40 keV. Exposure rates up to 2 R/h can be measured. Alpha particles with energies above 4 MeV are detected by a ZnS(Ag) crystal 3.82 cm in diameter. This last scintillation probe shows no sensitivity to photons up to 1 MeV in energy, even though the exposure rate is as high as 2000 R/h. Plastic scintillation materials are also in use for various specialized purposes and their use is growing as more materials are developed. Two typical scintillation type monitors are shown in Figure 5.4.

5.1.6 Proportional Survey Monitors

Proportional counters are used mainly as special purpose survey instruments. There are two important categories where proportional counters are particularly useful: to detect neutrons and to detect low-energy beta particles and photons. Typical instruments for both applications will be briefly discussed.

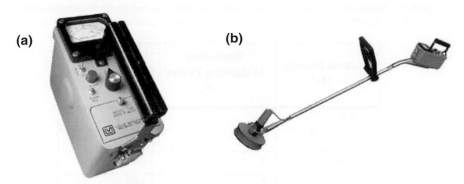

(a)   **(b)**

Figure 5.4: A typical scintillation survey monitor with an internal NaI(Tl) crystal is shown in (a). A plastic scintillation detector mounted on a rod with a monitor on the other end is shown in (b). (Courtesy of Ludlum Measurements, Inc.)

5.1.6.1 Neutron Monitoring

A typical neutron detecting proportional counter is shown in Figure 5.5. This instrument is light and rugged as well as battery-powered, hence it is an ideal unit for locating neutron leakage in inaccessible areas. The detector consists of a polyethylene cylinder containing $^{10}BF_3$ (boron trifluoride) filling gas at approximately 27 kPa pressure and neutron energy compensating materials. The energy detection range spans that from thermal neutrons (0.025 eV) to 15 MeV with a nominal neutron sensitivity of 2000 counts per 10 mSv. Since the quality factor for neutrons changes as the energy of the neutrons change, Table 5.1 gives a summary of neutron quality factors versus energy, along with the number of neutrons per centimeter squared per second (neutron flux density) at that energy which in a period of 40 hours results in an equivalent dose of 1 mSv.

Figure 5.5: typical proportional counter detector with measuring apparatus that together compose a portable high sensitivity neutron counter. There is no check source included with this instrument. (Courtesy of Ludlum Measurements, Inc.)

Table 5.1: Radiation weighting factors, w_R, and values of neutron flux density (number of neutrons per centimeter squared per second) which, in a period of 40 hours result in an equivalent dose of 1 mSv.

Neutron Energy — MeV	Radiation Weighting Factor — w_r	Neutron Flux Density[b] $cm^{-2} \cdot s^{-1}$
$2.5 \cdot 10^{-8}$	5	680
$1 \cdot 10^{-6}$	5	560
$1 \cdot 10^{-4}$	5	580
$1 \cdot 10^{-3}$	5	680
$1 \cdot 10^{-2}$	10	700
$1 \cdot 10^{-1}$	10	115
$5 \cdot 10^{-1}$	20	27
1	20	19
2.5	10	20
5	10	16
7	10	17
14	10	12
20	10	11
40	5	10
60	5	11
100	5	14
200	5	13
300	5	11
400	5	10

[b] Neutron Flux Density values from NCRP Report Number 39.

Proportional counters used as neutron detectors typically have a measuring apparatus similar to that described for G-M tubes and scintillation probes. The measuring apparatus shown can be used in count rate or integral mode and the results obtained in Sv/h, rem/h, Sv, rem, counts per minute, counts per second, or total counts. Additionally, it can be equipped with an audible speaker or earphone; it can be removed from the detector for remote monitoring and/or plugged into the AC current outlet for continual monitoring. Furthermore, the measuring apparatus can be computer controlled directly through a RS232 connection or remotely using an infrared link.

5.1.6.2 Low-Energy Particle Monitoring

A second major use of proportional survey monitors is for detecting small concentrations of low-energy beta particles and photons. A proportional detector with a large window area (100 cm^2 or 200 cm^2) for radiation acceptance has been developed for monitoring hands, shoes, clothing, including laundry, and surfaces such as counter tops and floors, for radionuclides contamination. As can be seen from Figure 5.6, these units use the same measuring apparatus as the neutron detector, and so can be continuously operated as well as both remotely and computer controlled.

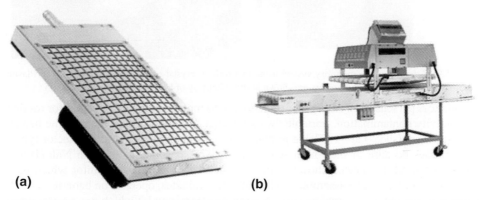

(a)   **(b)**

Figure 5.6:   A xenon filled gas proportional counter for measuring large areas for beta and photon contamination is shown in (a). A P-10 (10% methane; 90% argon) filled gas proportional counter for measuring laundry contaminated with alpha and beta particles is shown in (b). (Courtesy of Ludlum Measurements, Inc.)

5.1.7 Survey Instruments for Container Monitoring and Emergency Situations

Portal monitors which consists of either gas-filled or scintillation monitors are another important category of survey instruments. Portal monitors can be used to monitor waste disposal - all the garbage containers are passed through a portal monitor before being sent for disposal. Portal monitors are also used to monitor vehicles such as cars, trucks, and ships to detect the presence of radioactive materials, and can also be used to monitor a person or group of persons who may have been contaminated in a radioactive accident or other situation. The detectors in these systems can be G-M tubes, NaI(Tl) crystal or plastic scintillators. Plastic scintillation devices can be fabricated in many shapes and sizes and are useful for monitoring over a large area such a corridor, a roadway or harbor entrance. The monitoring systems are digital, usually computer controlled, and have alarm systems that can be set to respond to specific levels of radioactivity. Figure 5.7 shows in (a) lead shielded NaI(Tl) crystal detectors used to determine that hospital or industrial waste is free from any radioactive contamination, and in (b) plastic scintillation detectors being used to monitor vehicular traffic.

Figure 5.7: A typical waste monitory system using two NaI(Tl) crystal scintillators is shown is (a). Plastic scintillators used as vehicular monitors are shown in (b). (Courtesy of Ludlum Measurements, Inc.)

To monitors persons in an emergency situation, or to monitor radiation workers for routine radioactivity contamination, portable "walk through" monitors which monitor the hands, body, and feet (shoes) are available in many shapes and sizes and several detector types. Geiger tubes because of their high sensitivity as well as scintillators, both NaI(Tl) and plastic, are used. Figure 5.8 shows in (a) one such "walk through" monitor which uses plastic scintillants, can be assembled in less than five minutes, operates on batteries as well as line current and has a carrying case. A portable emergency kit which has several types of detectors which can be used with the monitor and can be brought to the scene of any disaster by first responders is shown in (b). These types of units are very useful in routine as well as in emergency situations.

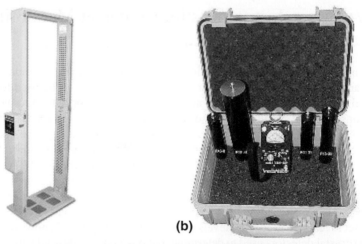

(a)   **(b)**

Figure 5.8: A portal monitor through which a person can walk is shown in (a). This monitor consists of plastic scintillation detectors. A first responders kit which has several interchangeable detectors is shown in (b). (Courtesy of Ludlum Measurements Inc. (a) and Technical Associates (b).)

5.2 Calibration Instruments for Radiation Sources

There are a number of instruments, based on ion chamber or semi-conductor detectors, which are used to periodically check the radiation output of diagnostic and oncologic X-ray generators (including dental and veterinary units) for compliance with the Nationwide Evaluation of X-ray Trends (NEXT) in the United States and other X-ray compliance programs. In diagnostic applications, especially in the case of fluoroscopic, mammographic, dental and veterinary units, it is important to know the total X-ray tube output and the value of leakage and scatter radiation since the radiation worker is often in the room with the patient during such procedures. Since radiation oncology generators deliver a very high absorbed dose rate to the patient, it is important to know this precisely. In general, all X-ray units and oncology sources should be checked at least twice yearly to determine the quality and quantity of radiation generation and to check the timing circuit.

5.2.1 Precision Electrometer/Dosemeter

This class of instruments is designed to be very reliable and accurate for measurements in the field of health protection, radiation physics, and industrial radiography. The energy dependence characteristics are excellent. These instruments can be used to calibrate X-ray beam output to determine total exposure at a given distance under given conditions of kilovolts, milliamperes and filtration. One representative instrument designed for routine day-to-day measurement, is illustrated in Figure 5.9. A wide selection of chambers can used with this instrument, thus offering a high degree of flexibility for various measurement applications. These applications include measuring electron and photon doses, electric charge or current, ^{60}Co units, and other radiation sources. A high or low range is switch selectable which can be used to extend the range of measurements by a factor of ten, maintaining linearity within +/-0.1% at a resolution of +/-0.005%.

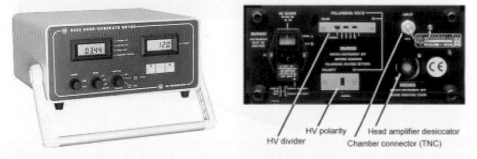

HV polarity   Head amplifier desiccator
HV divider   Chamber connector (TNC)

Figure 5.9: A typical combination dose and dose rate meter used to measure beam output from various electron and photon sources as well as charge or current. The electrometer is shown in (a) and close up the rear of the instrument is shown in (b). (Courtesy of Thermo Fisher Scientific, Inc.)

Typical ionization chambers used with this class of electrometers, generally consist of air-equivalent material (usually Bakelite or nylon). Since it is desirable for all the ion pairs to be produced exclusively in the air filling the chamber, different size chambers with different wall thicknesses are needed for photons of different energies. At present, chambers are available in four major categories (Table 5.2). Within each category, various chambers are available for different radiation source strengths. These chambers generally have a nominal accuracy rating of within ± 5 to ± 10% provided they are properly used.

Table 5.2: Four major categories of ion chambers

Type	Photon Energy — keV	Beta Energy Cut-off — MeV
Low Energy	6 to 35	0.066
Medium Energy	30 to 660	0.620
High Energy	250 to 1300	1.310
Skin Equivalent	30 to 250	0.070

5.2.2 Non-invasive Quality Assurance Test System

Many ion chamber monitor systems are available to determine exposure reproducibility, mA and time linearity, beam quality (HVL—see Chapter 7), exposure time accuracy, tabletop exposure rate, and kVp for X-ray generators in the diagnostic, dental, veterinary, industrial, and oncologic range. For example, Figure 5.10 illustrates how time, kVp, and exposure (or exposure rate) for various radiographic and therapeutic units can be determined simultaneously with a single exposure.

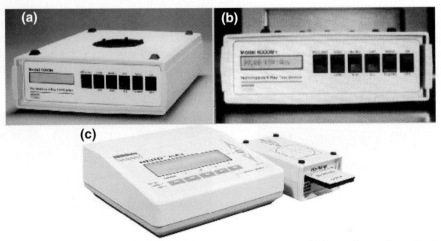

Figure 5.10: A non-invasive fully digital quality assurance test system for fast and easy determination of exposure, exposure time, kVp maximum, kVp effective, and kVp average for diagnostic, radiographic, or fluoroscopic generators is shown in (a) with a close-up front view in (b). Another instrument, shown in (c), measures kV maximum, kV effective, kVp average, time, exposure or exposure rate, mA or mAs, and other properties, offers the possibility to capture data from difficult machines. (Courtesy of Fluke Biomedical.)

These instruments usually have the measuring apparatus built into the device and provide an RS232 or wireless connection for a computer interface. Optional software which includes Excel spreadsheets for various types of calibrations are also available. The instrument shown in Figure 5.10(a) can be placed in the x-ray beam to make one exposure. The digital display serially shows kVp maximum, kVp average, kVp effective, effective dose, and time. The instrument then automatically resets for the next exposure. A CsI photodiode pair provide the kVp measurements through five user-selectable filter pairs. This ensures optimum accuracy over the entire diagnostic range with minimum filtration dependence. Exposure measurements are made with a parallel plate ionization chamber located above the filter wheel. Exposure time is measured with quartz crystal accuracy. For even greater flexibility, a variety of external ion chambers are available for specific generators, namely: radiographic, mammographic, scatter, image intensifier, computed tomography (CT), and CT high sensitivity for multislice systems. The device shown in Figure 5.10(c) provides evaluation of pulsed fluoroscopic, cine, CT, portable, mammographic, dental, radiographic, fluoroscopic, low, medium, and high frequency machines. It features nine operating modes and also provides external ion chambers for special purpose measurements. The separate detector unit can used to test generators to which access is generally difficult. There is many similar test instrument available.

A test system meant for daily use is pictured in Figure 5.11. Based on an ionization chamber detector, this is a durable, efficient, portable battery operated, electrometer for routine therapy beam dosimetry, electrometer capable of measuring electrons, x-rays and gamma rays from linear accelerators, cobalt 60 units, and other radiation sources. The measurement process is highly automated. This instrument can also control the precise dose delivered to the patient during treatment.

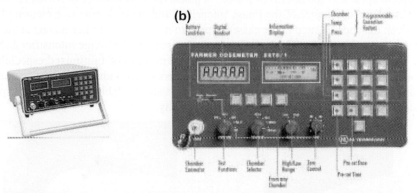

Figure 5.11: Digital constancy meter. The parallel-plate ion chamber detectors provide superior stability and optimal geometry for both electron and photon detection. A battery-portable digital constancy electrometer which has long-term stability and precision is shown in (a). A close-up of the front panel is shown in (b). This instrument can be used in routine measurement of electrons, X-rays and gamma rays from linear accelerators, cobalt 60 units, and other radiation sources to standardize radiation output before treatment. It can also control the precise dose delivered to the patient during treatment. Portable and rugged - offering excellent long term stability, it measures up to the day to day routines with the precision of a secondary standard. (Courtesy of Thermo Fisher Scientific, Inc.)

The use of this class of instruments is generally very simple: the detection device is placed on the treatment table, the exposure (using consistent geometry, field size and distance) is made, the result is read, recorded and compared with earlier measurements.This latter is usually done automatically under computer control. Hence, the mandated exposure and energy constancy checks can be done in a streamlined fashion.

5.2.3 Patient Air Kerma Monitors

One method of determining directly patient air kerma from diagnostic units is to use flat ion chambers or and metal-oxide-semiconductor field-effect transistor (MOSFET) devices transparent to visible light and X-ray photons mounted directly on the collimator face of the X-ray generator or on the patient's skin. Remote readout units can be located up to 15 m from the detector. Measurements are obtained in mGy·cm^2. Thus skin entrance air kerma can be calculated by dividing the reading by the field size. A built-in timer permits measuring of the integral time that the X-ray generator is on, yielding the total number of seconds of air kerma. Thus μGy/mAs can be calculated for both radiographic, fluoroscopic, and mammographic procedures. For certain applications, it might be desirable to have two or three ion chambers located in different places. These can be all connected to the remote reader via a distribution box, but only one chamber can be used at a time. A MOSFET device based system is shown in Figure 5.12 for use on a patient.

Additionally, ion chambers and MOSFET devices capable of detecting photons and electrons with remote measuring apparatus and readout units are used to measure the input air kerma levels directly at the entrance of the image intensifier or fluorescent screen. With a mechanical apparatus (phantom) or patient in place, the ion chamber or MOSFET device can be placed at the entrance of the image intensifier and the air kerma rate measured. The chamber can then be mounted in the cassette space to measure the air kerma rate at the back face of the image intensifier. The X-ray generator output and the image intensifier gain can then be adjusted for optimal conditions. Patient air kerma can be minimized especially by the avoidance of retakes. A check source is usually provided with this type of instrument to ensure its accurate functioning. Similar instruments are available for monitoring special units such those used in mammography or angiography.

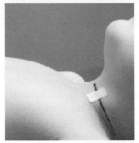 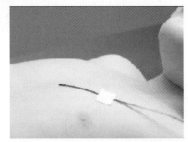

Figure 5.12: A small MOSFET skin entrance air kerma monitor shown at different positions on pediatric patient. The entrance skin dose can be measured remotely and thus the X-ray or other generator output can be adjusted as necessary. (Courtesy of Best Medical Canada, LTD.)

5.2.4 Calibration of Radioactive Sources

Another important use of ionization chambers is as a "dose calibrator". The purpose of this instrument, which is pictured in Figure 5.13, is to confirm exactly the amount of radioactivity in a given source. Whenever a vial of radioactive material is received or produced, e.g., by eluting a generator, or a quantity of such material is prepared for administration, this instrument is used to verify that the calculated amount of radioactivity is indeed what is present in the vial or syringe. This instrument is especially important when radioactivity in any form is to be administered to a person for research, diagnosis or therapy.

In general, the ionization potential is maintained in the chamber by an electronic circuit run from the wall outlet. In this application the instrument is used in current reading, mean-level mode, i.e., the voltage in the ionization chamber is kept at a specific level by constantly draining electrons off the anode. The number of electrons captured per second, (the total charge per second, or the current), is then used to calculate the number of counts per second or the activity of the source. There is generally a potentiometer supplied with this instrument which adjusts the resistance in the circuit (see Chapter 2, Figure 2.1) to maintain an optimal voltage level in the chamber consistent with the energy(ies) of the decay product(s) of the radionuclide being calibrated. There are also several scales for reading different levels of activity. These instruments can be computer controlled and interfaced to a printer for a permanent record.

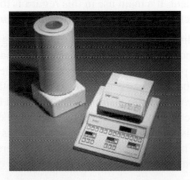

Figure 5.13: A typical standard ionization chamber configured to be a dose calibrator. This instrument is fully microprocessor controlled. The display can be read in curies or becquerels and there is self-diagnostic software. The unit comes with a dot matrix printer and an RS-232 bi-directional serial communications. Different models are available as well as one that is suitable for PET applications.(Courtesy of Biodex Medical Systems, Inc., which distributes the AtomicLab Dose Calibrator.)

The detector is usually air filled and is kept at normal temperature and pressure. The physical instrument is a cylindrical shaped chamber having an opening with a diameter of about 7.5 cm (3 in) and a depth of 25 cm (10 in) in which a small vial or syringe containing radioactive material can be placed. When the instrument is used for verifying the activity administered to patients, there are strict regulations regarding instrument calibration.

Each day a constancy check is performed by using a ^{137}Cs and/or other calibrated standard source(s) which are available in the United States from the National Institute for Science and Technology (NIST — formerly the National Bureau of Standards) in Gaithersburg, MD. Reading are taken on each of the potentiometer settings, recorded and compared with previous results. The linearity of the instrument is checked quarterly by counting the same radioactive source at frequent intervals for a specific period of time. For example, a ^{99m}Tc source might be counted each hour during a working day and again the next day. The readings obtained from the dose calibrator are plotted along with the theoretical values for radioactive decay (see Chapter 11) of the source used. An accuracy test is performed once a year, by reading a standard source of activity on the lowest scale setting and comparing the value obtained with the theoretical (calculated) value.

When the instrument is first received or when it is returned from repair, it must be checked for constant geometric response, i.e., the reading should not vary for small increments in volume of radioactive material in the vial or syringe. To check the geometric response of a dose calibrator, a vial is counted with the same radioactive content, but with the volume varied from 0.5 to 20.5 cm^3. A syringe is counted, again with the same radioactive content and the volume varied from 0.1 to 6 cm^3. The total capacity of both the vial and syringe used should be recorded.

Because of the need to set the potentiometer, the dose calibrator implicitly verifies the radioactive nuclide present. In addition it acts as a check on the theoretical calculations used to ascertain the radioactivity which is present. Hence, it can be thought of as a radiation protection instrument. It's main use in this regard is to protect the patient or other person receiving the radioactive material from an incorrect administration.

Traditional dose calibrators are optimized for counting gamma ray photons in the 100 to 300 keV region and do not give an accurate estimate of radioactivity for the annihilations photons from positron emitters or for beta particles. Due to an increase in the use of both positron and beta emitting radionuclides in both diagnostic and radiation oncology applications, dose calibration instruments optimized for measuring activity from these radionuclides have been developed. In general, instruments suitable for measuring annihilation photons can also be used for measuring photons of lower energies and are based on ion chambers. There is one company which offers a unit for counting beta particles only which is based on a solid state NaI(Tl) crystal as opposed to an ion chamber.

5.2.5 Calibration of Survey Monitors

All survey monitors are calibrated by the manufacturer at the time of production. Nevertheless, changes in the characteristics of the individual components of instruments may cause a change in the instrument response. It is, therefore, essential that all survey monitors be calibrated periodically to ensure proper reading. This calibration is especially important for G-M tubes because of their energy dependence characteristics. It is recommended that Geiger-Müller survey monitors be calibrated quarterly and after each battery exchange. They should be calibrated as often as necessary to meet the regulations of the NRC, agreement states, or other local regulations.

Photon detecting survey instruments such as G-M tubes and scintillation counters are frequently calibrated using a fixed geometric arrangement between the radiation source and the detector. The radiation sources most generally used are a known quantity of radiation of the same type and energy as the radiation to be monitored. Frequently, needlelike sources of ^{137}Cs are used as photon sources. The calibration technique should approximate the conditions under which the instrument is used. These counters feature a selector switch that allows the user to set the monitor to read a suitable maximum counting rate. It is important to calibrate the meter over the entire scale in each count-rate setting to compensate for any counting-rate dependence. The distance between the detector and the radiation source is varied in order to obtain different count rates. This effect of distance is discussed more fully in Chapter 7. A typical calibration instrument is shown in Figure 5.14.

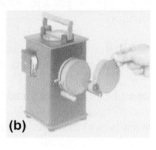

Figure 5.14: Gamma Survey Instrument Calibrator. (a) All three attenuators in place; (b) One attenuator removed from radiation path. (Courtesy of Fluke Biomedical.)

This device contains a shielded source of ^{137}Cs kept in either the stored or exposed position. In the fully-shielded stored position, the radiation at the container's surface the exposure is rated as less than 2 R/h and at 1 m away is less than 0.1 R/h. The source is moved from stored to exposed position by raising the control rod. The radiation field can be varied by moving the three built-in attenuators. These permit calibration of the three monitor scales, each at 20 and 80% of full scale, with only one source to chamber distance for every measurement.

Ion chambers, which are used to determine exposure or air kerma rates values from radiation generators, are sent to special centers for calibration. In the United States the calibration laboratory is NIST. Regional calibration laboratories have been approved by the American Association of Physicists in Medicine (AAPM). They are (as of November 2006):

    M.D. Anderson Hospital, Houston, TX.

    K & S Associates, Nashville, TN.

    University of Wisconsin, Madison, WI.

5.3 Review Questions

1. List some of the characteristics of an effective survey instrument. Can any one monitor satisfy all of these requirements?

2. What is meant by energy dependence? Why does it occur?

3. What type of survey instrument(s) show poor energy dependence characteristics?

4. Ionization chamber monitors are used to detect what type(s) of radiation?

5. What type of gas and what pressure are usually used in ionization chamber monitors?

6. If an ionization chamber is used for detecting alpha and beta particles and photons, how can the alpha and beta particles be sorted out?

7. Describe a typical ionization chamber monitor and its use.

8. What is (are) the major use(s) of proportional counters as survey monitors?

9. A laboratory worker monitors herself and her working environment with a proportional counter. Explain why this is a very good practice. Since this is a proportional counter monitor, what type of radiation is expected to be present?

10. G-M tube monitors are used to detect what type(s) of radiation? Are these instruments usually portable, i.e., battery operated?

11. What type of survey instrument is used as an alarm system? Describe briefly how this alarm system can be operated.

12. What are the advantages of a scintillation detector as a monitoring device? What are its disadvantages?

13. What type of scintillation material is commonly used in scintillation survey detectors?

14. What are the major uses of portal monitors? What type of detectors are usually used in portal monitor?

15. Should a portal monitor be extremely sensitive? If so, why?

16. If it is suspected that a person or group of persons has be exposed to ionizing radiation, what is a good monitor to use to check for internal radioactivity?

17. In an emergency situation, what types of monitors may be use to check whether or not ionizing radiation is present? What types of detectors are best to use?

18. Is it necessary to calibrate radiation sources such as X-ray generators? What characteristics of these radiation sources should be checked?

19. Describe two types of survey instruments used to check radiation sources.

20. What instruments can be used to check the daily calibrator of high energy generator for oncology?

21. What is a dose calibrator? How is it used? Why is it important as a radiation safety instrument?

22. What types of monitors may be used in to check the entrance skin dose for a particular procedure?

23. How are survey instruments such as G-M tubes and scintillation detectors calibrated? How are ionization chamber monitors calibrated?

5.4 Problems

1. A "cutie pie" ionization chamber monitor gives an air kerma rate of 5 cGy/h with the cap removed. It gives an air kerma rate of 200 cGy/h when the cap is affixed to the ion chamber. What can be deduced regarding the composition of the radiation?

2. A G-M detector has an audible alarm set to trip when the radiation level detected exceeds 10,000 cps. A ^{60}Co source, 80 kBq in strength, has been accidentally left unshielded. If the G-M tube detects the radiation with an efficiency of 20%, will the alarm sound?

3. A scintillation detector made of ZnS(Ag) is used to check for alpha particle contamination from ^{226}Ra. If the detector is 90% efficient and shows a count rate of 500 cps, what is the activity (in Bq) of the ^{226}Ra source?

4. A proportional counter monitoring for neutrons, shows an equivalent thermal neutron response of 10,000 thermal neutrons per centimeter squared per second. If a person were to be in this field for 10 s, what would be the approximate equivalent dose received?

5. A technologist is accidentally and unknowingly contaminated with 10 MBq of ^{14}C. When leaving the leaving the laboratory, the technologist monitors his/her hands with the wall-mounted proportional counter. If the monitor is 80% efficient and can register a maximum reading 3000 cps, will the meter deflect full scale?

6. An electrometer ion chamber monitor is used to study the X-ray output of a fluoroscopy machine. At a given kVp, the mAs is set to give an air kerma value of 10 cGy in 3 s. The ion chamber monitor registers a total air kerma of 40 cGy. By how much is the timing circuit off? How could it be determined that both the kVp and the mA are correct?

Chapter 6

PERSONNEL MONITORS

Determination of a person's exposure (air kerma) to ionizing radiation is an important function of radiation protection in general and is particularly necessary for both radiation workers and for others such as secretaries and housekeepers, who have access to areas where radiation may be present. Additionally, nurses and other persons who may care for patients who have received radioactive material need to be monitored. In this chapter we describe the most common types of personnel monitors used for evaluating exposure due to photons which easily penetrate all parts of the body and are virtually undetectable unless an external device is used. The monitors described also evaluate the effective/equivalent dose from electrons. If it is possible that a radiation worker will be exposed to neutrons or to heavy charged particles, the area rather than the person is usually surveyed.

Thus, all personnel who may be exposed to even very low levels of radiation need to be monitored. This monitoring can be accomplished in a number of ways. The particular method chosen depends on the type and level of radiation to which the worker is most frequently exposed, as well as the cost and dependability of the monitoring method. It must also be kept in mind that the actual reading on the personal monitor does not correspond directly to the whole-body dose. This must be calculated based on a model of exposure for the activity in which the individual is engaged. In the usual case in which the whole-body exposure is fairly uniform and is mostly due to photons and high-energy electrons, the exposure value obtained on the personnel monitor is a fairly accurate estimate of the whole-body effective dose. However, when the exposure is highly nonuniform in distribution or type of radiation, a more detailed assessment must be made.

6.1 General Characteristics

A personnel monitor should be reasonably accurate and reliable. It should also be sensitive to the type(s) of radiation being monitored. Since every radiation worker should have one, it should be inexpensive, portable (ideally, clipped to some article of clothing), and easy to understand (require no technical knowledge of the user). The last two are crucial because they allow radiation or other workers to go about their ordinary tasks without being conscious of the personnel monitor. Hence a worker's ordinary exposure can be evaluated in their working environment.

Personnel monitoring devices can be divided into essentially four different categories:

1. those that are used for detecting low levels of radiation and are designed to be used over a specified period of time (such as a week or a month), for example, X-ray or nuclear medicine technologists, radiologists, nuclear power plant engineers, and technologists; or for monitoring persons who may be incidentally exposed to low radiation levels, e.g., nurses and housekeeping staff who work in areas where radiation is used;

2. those that are designed to be read whenever the need arises, such as for monitoring radioactive patients and their caregivers; or for radiation workers such as radiation therapy technologists who may be exposed to high levels of radiation;

3. those that give an audible and visual warning in response to an unexpected radiation level which has reached a pre-set threshold; for example, near a fluoroscopy or angiography unit, or in a nuclear power plant.

4. those that can be used when airborne radioactivity may be present.

Other devices are also used for monitoring. However, they are not commonly used for monitoring in the radiation protection field so they will not be discussed.

6.2 Delayed Readout Personnel Monitors

These types of monitors are intended to be worn for a period of time. The most popular types of devices are those made of film or those composed of inorganic crystal scintillators which must undergo some external process before the trapped electrons are released. Generally speaking, employers who use these types of devices for personnel monitoring, contract with a separate concern to supply the devices and to collect and evaluate used ones. This contractor then provides readings of the devices to the employer, who in turn, passes this information on to their employees. This multistep procedure takes time, so the results often arrive one month or more after the period during which the monitor was worn.

It is common in many laboratories and clinics to have one extra monitoring device placed in some generally unexposed portion of the room or department. This monitor, viewed as a control, is then evaluated together with those of the employees. It serves to separate exposure due to background levels of radiation from exposure an individual worker may acquire as a result of his/her work habits. In this way high background levels can be detected and eliminated, and careless workers can be admonished.

Another advantage of these devices is that they can be molded into various shapes and sizes. For example, radiation workers who regularly use radionuclides usually wear a ring monitor on one finger in addition to that clipped to their clothing. The ring device then specifically monitors the hands which are often in close contact with photon-emitting substances as tasks such as eluting the generator, preparing and administering necessary radiopharmaceuticals are performed. Other special-purpose devices, such as those worn on the wrist and ankle, also exist for monitoring particular areas of the body.

6.2.1 Film Badges

One of the more popular types of personnel monitoring devices is the film badge (Figure 6.1). The film itself consists of a photographic emulsion mounted in plastic. This plastic-mounted film is wrapped in thin, light-tight paper, sandwich fashion and placed in a plastic holder. The holder clips onto the wearer's clothing, enabling the film badge to give a measure of total, or whole-body, exposure. The plastic case usually contains a cutaway portion to allow the entrance of beta particles or other electrons. It also contains three small metallic filters - usually copper, cadmium, and aluminum - placed in different portions of the case to help distinguish among higher energy photons. Each of the metals attenuates photons of different energy values. This packaging has been devised so that the film badge can measure the exposure which the wearer has encountered and can also help distinguish the type of radiation to which he/she has been exposed. The outside of the film wrapper has the name, date of issue, and identification number of the wearer imprinted on it.

(a)   **(b)**

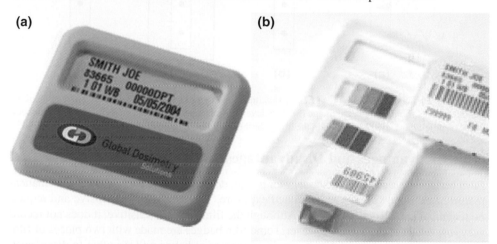

Figure 6.1: The Film Badge: The assembled film packet and plastic case is shown in (a); A cut-away view of the plastic case exposing metallic filters and hole for the passage of beta particles is shown in (b). (Courtesy of Global Dosimetry Solutions, Inc.)

The film badge operates in the following way. When a photon interacts with the film emulsion, it undergoes a photoelectric or Compton interaction, thus releasing an electron which deposits its energy. This energy deposition alters some of the silver halide grains that comprise the film emulsion. These altered grains then respond differently to the reducing agents, known as developers; i.e., those grains affected by radiation are reduced to metallic silver whereas the grains not so affected are not. This developable state produced in the emulsion by the action of radiation is called the latent image. The optical density of developed film, i.e., its degree of darkening, is proportional to the exposure of the film (Figure 6.2). The relationship between exposure and optical density is determined by comparison with films exposed to known amounts of radiation of the same energy, since optical density produced by a given exposure of radiation is strongly dependent upon the energy of that radiation. It should be noted that photographic emulsions have a specific energy response, i.e., are most sensitive to photons in a particular energy range. It is, therefore, useful to include metallic filters, which by attenuating photons in specific energy ranges, create electrons (via the photoelectron and Compton interactions) that are then detected by the film. By looking at the positions of the resulting density variation of the film, the reader (based upon a film densitometer) is able to estimate the energy as well as the amount of radiation to which the film was exposed.

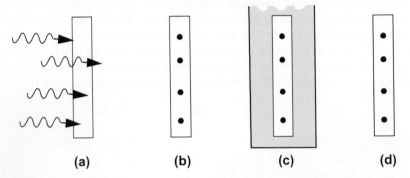

(a)   (b)   (c)   (d)

Figure 6.2: Radiation and Film: (a) shows the radiation passing through the emulsion activating some of the grains and thus forming the latent image (b). The development stage (c) reduces the emulsion and produces, in (d), visible grains. These last contain the image or the record of passage of the ionizing particles.

6.2.1.1 Advantages and Disadvantages

Film badges are a popular personnel monitoring device because they provide a permanent record of each individual's accumulated exposure. They are also inexpensive and require no technical knowledge of the user. Although the film badge is sensitive, it does not record very low levels of exposure. However, some film badges are made with two pieces of film instead of one, one piece to detect very low levels of radiation and the other to detect even very high levels of radiation. In general, such an elaborate badge is not necessary because the lower levels of radiation not recorded on the film are insignificant, and very high levels of exposure are better monitored in other ways. As an example, radiographers whose main work involves exposure to ionizing radiation at oncologic levels very often wear an ion-chamber pocket dosimeter as well as a film badge.

Whether the film is developed in-house or by an outside contractor, its advantage is the permanent record thus provided. If necessary, film can be reread. This re-reading capability offers a distinct advantage because a mistake could be made, particularly if the reading is too high. (The limitations on effective/equivalent dose values for various types of workers were discussed in Chapter 4.) The film badge is the only personnel monitoring device that has this advantage. It should be mentioned, however, that film is not accurate for any exact measurement purposes. Film can give only an indication of radiation level. It also has low reproducibility.

Because the film badge is essentially a photographic device, two basic precautions apply:

1. Heat affects film; hence the badge should not be exposed to sunlight or stored in a warm place, such as an automobile (because of the greenhouse effect).

2. The latent image begins to fade after a time, so a piece of film should not be used for more than one or two months before developing.

General practice dictates that film badges be replaced biweekly or monthly. When radiation levels are low, a monthly change of film badge is sufficient. Cost, as well as insensitivity of film to very low levels of radiation, prohibits changing of film badges at more frequent intervals.

6.2.2 Thermoluminescent Dosimeters (TLDs)

These personnel monitors can be used in the same way as film badges. They are common for whole-body monitoring as well as for special types of monitoring, such as that described for the hands. Since they may be affixed anywhere in the treatment field and may be attached to a phantom or directly to a patient, they are often used in radiation oncology applications (in particular to verify the treatment plan). Thermoluminescent dosimeters are excellent monitors and are enjoying a well-deserved popularity.

Substances that possess the property of thermoluminescence are nonmetallic crystalline solids, usually in powdered form. As was discussed in Chapter 2 with regard to solid state detectors, when electrons in a crystal insulator absorb energy, they move to the higher energy conduction band and all the missing electrons or holes migrate to the valence band. In a thermoluminescent crystal, these excited electrons get trapped in the higher energy state until the crystal is heated to a specific temperature, known as the curie temperature, at which the electrons return to the valence state, radiating the extra energy in the form of visible light photons. It is from this mechanism that their name, thermal (heat) luminescent (light-giving) dosimeters, is derived. Lithium fluoride (LiF), lithium tetraborate ($Li_2B_4O_7$), and manganese-activated calcium fluoride (CaF_2:Mn) are crystals commonly used for their thermoluminescent properties. The curie temperature of these substances is 845, 930, and 1423 °C respectively.

Personnel monitors that contain a thermoluminescent substance in powder or other solid-state form such as a ribbon, can be molded into a shape appropriate for the intended function. When a thermoluminescent substance is exposed to ionizing radiation, electrons in the crystalline structure of the material are excited to higher energy states. A certain number of these excited electrons get trapped in what are known as sensitivity centers. When the crystals are subsequently heated, the amount of light obtained is proportional to the energy absorbed by the crystal. If the crystal is then annealed properly, in other words, cooled slowly, allowing it to relieve stress in its lattice as the temperature is brought down, the crystal structure is not damaged and the crystal is ready for use again. LiF, however, must be recalibrated after each use, while CaF_2:Mn not only has a simpler annealing technique, but does not need recalibration, i.e., its sensitivity and stability are not dependent on prior excitation quantity or thermal history. The exposure range over which LiF is linear is smaller than that for CaF_2:Mn; however, although the latter is more energy dependent, it is sensitive to exposure values one tenth of the lowest exposure limit for LiF. $Li_2B_4O_7$ combines the good features of both of the above substances and may ultimately prove to be a better overall choice for personal monitors.

6.2.2.1 Advantages and Disadvantages

One advantage of thermoluminescent dosimeters is that they can be reused. It is relatively easy to determine an exposure value from such a device because all that is required is an oven and a light meter. Since the crystalline structure and energy levels of the substance used are well-known, a direct measurement of energy absorbed is obtained. Also, thermoluminescent dosimeters are not as sensitive to moderate heat as film. They do suffer from fading of information, but this can be overcome by various techniques. Furthermore, when compared to film, they have an increased sensitivity to a wide photon energy range. TLDs can be used for several weeks at a time without appreciable loss of stored energy. Most important, they require no technical knowledge for use.

Disadvantages of TLDs include higher cost than that for a film badge and, more significantly, a nonpermanent record of exposure. Once the TLD has been heated and annealed, the information is gone. Recently, some researchers have attempted to eliminate this reproducibility problem by setting aside some crystal grains. However, this is costly and the resulting material and data management are problematic.

6.2.3 Optically Stimulated Luminescence (OSL)

This is the newest advance in passive detection of ionizing photons and beta particles. Optically stimulated crystals have the same response to irradiation as TLDs and can be used similarly as the personnel radiation detectors. The difference is only in the readout mechanism. These new devices are usually activated to release the trapped electrons when they are irradiated with ultraviolet, blue-green, or infrared light. Quartz is an example of a naturally occurring crystal which can be activated in this way. Many artificial materials such as porcelain and carbon activated aluminum oxide (Al_2O_3:C) also exhibit this OSL property. These materials can retain a record of the accumulated radiation for a long period of time and so have been useful in determining immediate and long-term irradiation in such areas as the Japanese bomb sites and Chernobyl. Furthermore, Al_2O_3:C has also been made into dosimeters similar to TLDs and film badges. These are excellent monitors and in many applications are slowly replacing film and TLDs.

6.2.3.1 Advantages and Disadvantages

The advantages of OSL devices are similar to those of TLDs in that they can be reused, have an increased sensitivity to a wide range of photon and beta particle energies, and are not sensitive to moderate heat. Since the crystalline structure and energy levels of the substance used are well-known, a direct measurement of energy absorbed is obtained. These devices do not suffer from fading of information, and consequently can be used for up to year or more. Most important, they require no technical knowledge for use.

Disadvantages of OSLs include higher cost and, again, a nonpermanent record of exposure. Once the device is read, the information is gone.

6.3 Immediate Readout Personnel Monitors

The monitors discussed so far do not give the user a way to immediately determine his/her exposure. The active element (film, TLD, or OSL) must be read in a controlled environment. In addition to these monitors, solid state and ion chamber monitors have been designed so that exposure levels can be read at any time, either by the user or by some other designated person. These are meant to be worn on the clothing and the readout mechanism is simple and easy to use.

6.3.1 Solid State Devices

These simple devices are shaped like a small, flat object usually no more than 15 cm (6 in) in length. They usually have an LCD digital display and are capable of detecting photons (depending on the model) in the range of 0.02 to 1 MeV and indicate the integrated radiation dose equivalent in μSv. Typically they weigh about 50 g, use a lithium battery with a lifetime of two weeks to one month of continuous operation. No external charger or reader is necessary. Figure 6.3 shows a typical solid state personnel detector.

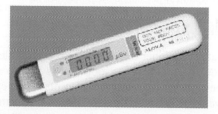

Figure 6.3: A typical solid state personnel monitor. It is small in size and weighs only 50 g. This monitor will detect gamma radiation from 0.02 to 1 MeV and has an operating temperature range from 0 to 45 °C. (Courtesy Cone Instruments, Inc.)

6.3.1.1 Advantages and Disadvantages

The advantage of these monitors lies in their simplicity of use, their long life time, and their direct reading capabilities. They are rugged and require no maintenance except for a simple battery change. They can easily worn by persons working around high or continuous X-ray sources (such as fluoroscopy or angiography units), for monitoring patients who are undergoing radioimmuniotherapy, for use in emergency kits, or in facilities such as nuclear power plants, for workers and visitors. The monitors reset with a simple on/off switch. The disadvantage is, of course, that they are expensive and provide no permanent record.

6.3.2 Ion Chamber Devices

These simple ion-chamber monitors are shaped like a large fountain pen and attach to the clothing (Figure 6.4). As with other ionization chambers, these monitors must be calibrated for the energy range of the radiation to be monitored as well for permissible amount of charge leakage per week for the chamber to remain in calibration. If the barrel of the chamber is made of aluminum, another correction factor, may also have to be employed. Two kinds of ion-chamber monitors are available. One is known as a direct-reading pocket monitor, because the person using it can read it at any time. The other is a condenser-type pocket chamber, which requires a separate device for both charging and obtaining results. A brief description of each type is given next.

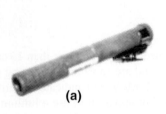

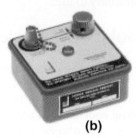

Figure 6.4: (a) A typical Ionization Chamber Personnel Monitor. (b) A device for recharging the ionization chamber monitor shown in (a). (Courtesy of Ludlum Measurements, Inc.)

6.3.2.1 Direct-Reading Monitors

This monitor is an ionization chamber containing two central electrodes, one of which is fixed. The other electrode is a quartz fiber loop that is free to move with respect to its mounting. The role of the external charging unit is to place like charges on the electrodes. This results in a repulsive force between the two, which forces the loop outward from the mount. The ionization produced in the chamber by external radiation, particularly photons, reduces the charge and allows the fiber to move toward its normal position. This action is similar to that of an electroscope, which is commonly used for measuring charge (see Glossary).

The direct-reading monitor is cylindrical in shape. One end is made of glass to allow light to enter; a transparent scale forms the other end. The instrument is read by simply holding it up to light. The quartz fiber casts a shadow on the scale which is calibrated in exposure units. Hence the user can determine exposure immediately.

This kind of ion chamber is particularly useful for monitoring personnel such as physicians and nurses who are caring for hospitalized patients receiving oncologic levels of radiation which involve either radioactive implants or large doses of unsealed radionuclides such as ^{131}I. In this case, each person entering the patient's room is required to wear a pocket chamber while in the room. A log is placed outside the patient's room. Each person logs the chamber's reading before entering the room, then again when leaving. Each person also logs his/her name, the date, and the amount of time spent in the patient's room.

The operation of direct-reading dosimeters should be checked periodically. To accomplish this, a hermetically-sealed ^{137}Cs source (usually about 3 MBq in strength) can be utilized. Calibrators exist for checking either one dosimeter unit at a time, or up to six units simultaneously.

6.3.2.2 Condenser-Type Pocket Chambers

The condenser-type pocket chamber has a central wire and a cylindrical electrode insulated from a Bakelite wall (Bakelite is used because it is light, hard, effective as an insulator, approximately air equivalent, and heat resistant). A charge is placed on the center electrode by means of an external charging unit. Ions formed in the chamber reduce the charge by an amount proportional to the radiation exposure. The condenser-type pocket chamber differs from the direct-reading type mainly in that the functions performed by the quartz fiber and scale mechanism are in an external unit. Consequently, the chamber must be read with a separate charger-reader similar to that shown in Figure 6.4.

6.3.2.3 Advantages and Disadvantages

The advantage of condenser-type pocket chambers over direct-reading ones is that they are lower in price. The condenser type is most often used for monitoring radiation oncology personnel, such as technologists and physicians, because of the high probability of being exposed to substantial amounts of radiation. Should the necessity arise, the amount of exposure incurred could be determined immediately. Both kinds of ion-chamber devices must be recharged daily since significant leakage of charge may occur over long periods of time. Both kinds are read immediately after use. In addition, ion-chamber monitors tend to be fragile rather than rugged.

The ion-chamber type of personnel monitor is more costly than either the thermoluminescent dosimeter or the film badge, with the condenser type of ion chamber being less expensive than the direct-reading kind. However, direct-reading ion-chamber personnel monitors can be read immediately. It does take a certain amount of skill to do the reading, but the user can easily gain this capability. The condenser type does not require any skill to use. Because they should be read daily, ion-chamber personnel monitors are not necessary for monitoring persons who may be exposed only to very low levels of radiation. As with thermoluminescent dosimeters, ion chambers leave no permanent record of the user's exposure. Once the unit has been read and re-charged, the recorded (or possibly not recorded) exposure value cannot be double-checked.

6.4 Personal Alarm Dosimeters

Personnel monitors, some of which are the size and shape of a cigarette pack, fit easily into a pocket or can be clipped elsewhere. These are meant mainly as warning devices. They can be set to respond with an audible signal to any level of photon radiation or only after some minimum level of radiation has been reached. The audible signal emitted is frequently in direct proportion to the amount of radiation detected. Some detectors have a digital display that can be part of the unit or can be read remotely, showing the accumulated exposure. This reading is not affected by turning the monitor off; it can be reset only by a separate, usually protected, switch.

6.4.1 Solid State Type Detectors

Solid state personal radiation detectors for use in gamma detection are also available. They are rugged yet lightweight, and come equipped with both an audible and visual alarm. They usually include a very long-lived lithium battery (typically 1500 hours in background radiation) and contain a low battery indicator for timely replacement or recharging. They are generally packaged in molded plastic with LED displays for alert conditions. They only give an indication that radiation beyond a user definable threshold level has been reached. These detectors are sensitive and rugged, thus they are capable of being used in the same manner as G-M detectors described below. A typical monitor is shown in Figure 6.5.

Figure 6.5: A typical solid state personal radiation detector monitor. It is small in size and weighs only about 30 g. This monitor will detect gamma radiation from 0.02 to 10 MeV and has an operating temperature range from -10 to 45 °C.(Courtesy Ludlum Measurements, Inc.)

6.4.1.1 Advantages and Disadvantages

In general, these detectors are rugged and very easy to use. The audible and visual alarms are a significant early warning of unusual levels of radiation. They are also relatively inexpensive. The disadvantage is that it gives no indication of the amount of radiation present or of the exposure of the individual using it. It is only an early warning device.

6.4.2 G-M Type Detectors

Because of the sensitivity of G-M detectors, these monitors are ideal for personnel working in areas which could become accidentally contaminated through leakage radiation of some sort, or for fluoroscopy technologists or radiologists who work with complex radiation generating sources whose characteristic values may change without warning and also for emergency workers. These monitors are very useful in ALARA programs. A check source should be used with these monitors if possible, and they should be calibrated regularly. Figure 6.6 shows sample of a G-M type detector.

Figure 6.6: A typical halogen-quenched G-M personal radiation detector monitor. It is light in weight and small enough to fit into a shirt pocket. This monitor detects energy alpha, beta, and photon radiation. (Courtesy Cone Instruments, Inc.)

6.4.2.1 Advantages and Disadvantages

G-M monitors give no permanent record of the exposure accumulated. The exposure, together with the time period over which it was acquired, must be recorded. The exposure value and accumulation time can be read at will, but once the instrument has been reset, the former reading is gone. These monitors are more expensive than ion-chamber personnel monitors, but they have the advantage of an audible warning signal and are much more rugged. G-M monitors of this type are usually used in conjunction with a film badge or TLD for accurate recording of personnel exposure.

6.5 Personal Air Samplers

If a radiation worker, emergency or first responder personnel may be exposed to aerosols, particulate airborne radiation, or even radionuclides such as the various forms of iodine which sublimate readily, then a personal air sampler might be worn. This usually consists of a unit that can be clipped onto the collar or belt, and which contains filter paper of various degrees of coarseness, perhaps charcoal impregnated, or a charcoal filter for detection of radioiodine only. Air is drawn in through the unit by a portable pump attached via a flexible hose. The pump is usually operated by a rechargeable battery, records the total air volume registered, and has an air-flow rate dependent upon the particular filter used. Each time the filter is removed for counting, the air volume is noted and reset. The frequency of filter replacement is necessarily a function of the radiation contamination for which it is being used. For example, if ^{131}I is the radiation contamination of interest, the filter should be removed and counted daily. Figure 6.7 shows a typical air sampler with self contained pump and separate readout unit.

(a)   **(b)**

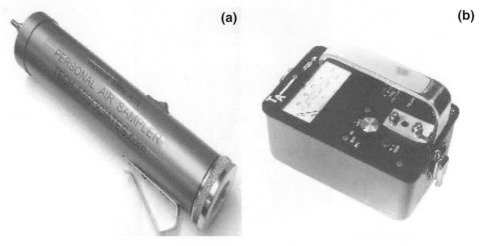

Figure 6.7: (a) Typical G-M type detector air sampler monitor with a self contained air pump unit. Integrated exposure information is provided by the separate readout unit (b). It is designed for industrial, hospital emergency personnel breathing air measurement systems, but can also be used to monitor fume hoods or stack effluents. (Courtesy of Technical Associates.)

6.5.1 Advantages and Disadvantages

These monitors are expensive, both to purchase and to operate (since a continuous supply of filters is needed), but when their use is indicated, e.g., when airborne radioactivity is likely to be or is suspected to be present, such monitors should be employed.

6.6 Other Personnel Monitoring Methods

For personnel who are exposed to internal radiation contamination from radionuclides, other monitoring methods may be employed. Whole-body counting is one method of externally determining an internally deposited radiation body burden. This method involves placing the person concerned in a heavily shielded room or in an apparatus known as a "shadow shield." The purpose of each is to reduce the background count level, particularly from cosmic radiation and building materials. The actual measurement is usually made with a scintillation detector, made of NaI(Tl). Large crystals, approximately 7 cm in diameter and 7 cm thick are used for whole-body evaluation. If only the thyroid is to be counted, especially when ^{125}I is to be detected, then a smaller crystal, usually 2.5 cm in diameter and 1 to 2 cm thick is used. In addition, urine analysis and blood count for both red and white cells should be routinely done. The analysis of feces, breath, nasal discharge, sputum, saliva, sweat, and hair can also be done, but are usually not done routinely.

6.7 Review Questions

1. What general characteristics would one like to have in a personal monitor?

2. Describe the usual film badge monitor. Can it be made into different shapes for special purpose monitoring?

3. How is the latent image on any film made? How are the metallic filters in a film badge related to the latent image formed?

4. Discuss the advantages and disadvantages of a film badge monitor.

5. How does a thermoluminescent (TLD) device work?

6. What are the advantages and disadvantages of a TLD as a personnel monitor?

7. When should a personal air sampler monitor be used? Would it be useful in the fluoroscopy suite? For a technologist using ^{125}I in animal experiments? In the control room of a nuclear power reactor? Give reasons for your answers.

8. What is the difference between direct-reading and condenser-type ion-chamber monitors? Briefly describe each type.

9. Would an ion-chamber personnel monitor be useful in the fluoroscopy suite? In the treatment area where a linear accelerator is used? Give reasons for your answers.

10. G-M personal alarm monitors are useful as warning devices. What are their disadvantages as the only source of personal monitoring?

11. Name some of the methods of assaying for internal contamination by radionuclides.

6.8 Problems

1. The monthly film badge report for a certain technologist shows an effective dose of 0.2 mSv of beta particles, 1.5 mSv of medium energy photons, and 0.3 mSv of high energy photons. What is the technologist's total effective dose for the month?

2. A nuclear medicine technologist wears a TLD ring monitor on one finger. What is the maximum reading the monitor can show each month if he/she is not to exceed the allowable effective/equivalent dose limits? Assume the irradiation is from photons and beta particles. Explain your answer.

3. A nurse is caring for a patient with a radioactive implant which decays via photons. A direct-reading dosimeter suitable for detecting the emitted photons is present for his/her use. The nurse logs in the dosimeter as reading 10 μGy before entering the patient's room and as reading 30 μGy after leaving the patient's room. Assuming the dosimeter to be 90% efficient, what is the total exposure? What is the effective dose?

4. A health physicist working in a nuclear power plant carries a personal alarm G-M detector set to trigger at 300 cps. What would be the minimal amount of activity (in Bq's) required to trigger the alarm if the G-M tube were 10% efficient (i.e., registered 10% of the photons reaching it)?

Chapter 7

PRACTICAL MEANS OF RADIATION PROTECTION

As evidence began to accumulate that ionizing radiation was inherently dangerous, practical methods of protection were developed. In essence, the three effective means of protection from ionizing radiation are: time, distance, and shielding. These principles are described in a theoretical way here. Later chapters detail how these principles are used in practical situations.

7.1 Time

The simplest protection from ionizing radiation is not to be where it exists. An important principle in dealing with radiation is to spend as little time as possible in the vicinity of radiation. Translated into practical terms, this principle has several implications.

As a first example, consider ordinary radiography. It is sometimes necessary to hold a patient in place, e.g., the patient may be quite young, elderly, or ill. The technologist or physician can best hold the patient; however, it would be much better for someone who is not a radiation worker to do the holding. If possible, a relative of the patient should be asked to help. This person is possibly being irradiated only once, whereas the technologist or physician constantly has the possibility of being irradiated. New York State has even gone so far as to make illegal the practice of persons occupationally exposed to radiation (e.g., technologist) holding patients during an examination. An institution is not even allowed to hire a person for that purpose or to utilize a limited number of persons regularly. Furthermore, the institution is required to monitor and maintain for inspection, exposure records for any person holding a patient. The ICRP recommends that those persons holding a patient be a relative of an age greater than the normal reproductive age, that they wear protective gloves and aprons, and ensure, as far as it is practical, that no part of the body, even if covered, be in the path of the useful beam.

When it is necessary to be exposed to radiation, a good rule of thumb is to develop efficient working habits and techniques. This measure applies particularly to radionuclide work and to fluoroscopy. For example, when the Mo-^{99m}Tc generator is eluted in a nuclear medical laboratory, it should be done skillfully, so that the person performing the elution has the shortest possible probability of being irradiated. The same applies to chemical preparations containing radionuclides, and it also applies to the preparation and administration to patients of radioactive materials. Similar situations could be cited for persons working in nuclear power plants where radioactive solid as well as gaseous wastes must be handled.

The less time spent in actual handling of radioactive material, the better. This rule is true even when such other protective methods as distance and shielding are used. The best exposure is no exposure. This same principle of skillful technique applies to fluoroscopy; the shorter the time during which the X-ray beam is turned on, the better. Thus the importance of respecting the ICRP principle "As Low As Reasonably Achievable" (ALARA) is evident. One effective method to help achieve this is to practice new protocols before using any actual radiation.

7.2 Distance

The most effective means of protection from ionizing radiation is a wise use of distance. A nice feature about distance is that particulate radiation has a very short range, even in air; as for photons, the number reaching a particular point in space per unit time decreases according to an inverse square law, as the photon beam spreads farther from the source.

This concept of an inverse square law requires further explanation. Picture a number of spheres of different sizes all centered on one spot, e.g., a golf ball, a tennis ball, a bowling ball, and a basketball, all having a common center and each inside the other (Figure 7.1). Now picture lines drawn from the center and intercepting the surface of all these balls. Since Figure 7.1 is drawn in two dimensions, the lines intercept the circumference or boundary of various circles. If a definite size block were then chosen and placed on the surface of each of the balls (or boundary of each of the circles), the number of lines intercepted by the block when placed on the smallest ball (or circle) would be greater than the number of lines intercepted by the block when placed on a bigger ball (or circle). As a matter of fact, the lines intercepted decrease in inverse proportion to the distance of the surface (or boundary) from the common center. Think of these lines as being representations of the path of photons.

From Figure 7.1 it is evident that the total number of lines that pass through the first circle also pass through the second and the third and the fourth and so on. This interception also occurs with three-dimensional objects. In this case the surface area of a sphere—such as a golf ball, tennis ball, or bowling ball—is proportional to the square of the radius:

$$\text{Surface area of sphere} \propto \text{radius squared}(r^2) \qquad (7.1)$$

Thus, the number of lines (which represent photons) per unit surface of the golf ball is proportional to the total number of lines divided by the radius of the golf ball squared:

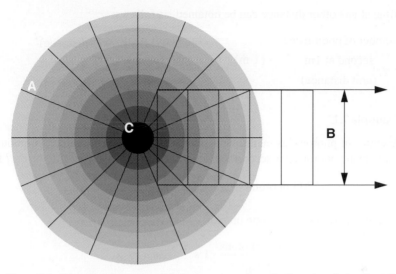

Figure 7.1: Diagram of concept of the inverse square law: the lines A radiating from the center C spread out as they go from inner circle to outer circle. If a constant length B is placed across the circumference (edge) of each circle, it is seen that fewer lines are intercepted by the length as it moves from the inner to the outer circles.

$$\text{Number of lines per unit surface area of golf ball} \propto \frac{\text{Total number of lines}}{r^2 \text{ golf ball}} \qquad (7.2)$$

The same rule applies for the number of lines per unit surface area on the basketball:

$$\text{Number of lines per unit surface area of basketball} \propto \frac{\text{Total number of lines}}{r^2 \text{ basketball}} \qquad (7.3)$$

Since the surface area of the basketball is larger, in that is has a much larger radius than the golf ball, and the total number of lines remains the same, the basketball must have fewer lines per unit surface area than the golf ball. Notice that the inverse square of the radius is used in the preceding calculation; thus the origin of the term, "inverse square law".

To express this relationship quantitatively, the following conversion is employed. The source of photons is treated as a point source located at the center of the sphere. In general, for reasonable distances and source sizes, this approximation is adequate. Then a reference point is picked: one unit distance away. This unit distance can be 1 meter, 1 centimeter, or whatever is the most convenient unit to use. In this text, 1 meter has been chosen for consistency. As a result, knowing the number of photons emitted per unit time at a distance of 1 meter (or any other unit distance) from the source, the number of photons per unit time at any other distance can be obtained, as follows:

$$
\begin{array}{l}
\text{Number of photons per} \\
\text{second at 1m} \\
\text{(unit distance)}
\end{array}
\cdot (1\ \text{m})^2 =
\begin{array}{l}
\text{Number of photons} \\
\text{per second} \\
\text{elsewhere}
\end{array}
\cdot
\begin{array}{l}
(\text{distance})^2 \\
\text{in meters} \\
\text{from source}
\end{array}
\qquad (7.4)
$$

Example 7.1

A source of photons, as measured at a distance of 1 meter, emits 1000 photons per second. How many photons per second will be measured at a distance of 5 meters? Hence:

$$
1000\ \text{photons} \cdot 1\ \text{s}^{-1} \cdot (1\ \text{m})^2 = x \cdot (5\ \text{m})^2
$$

Rearranging the above equation, one obtains:

$$
\frac{1000\ \text{photons}}{25\ \text{m}^2} \cdot 1\ \text{s}^{-1} \cdot 1\ \text{m}^2 = x
$$

which gives finally:

$$
40\,\frac{\text{photons}}{\text{s}} = x
$$

Three things are worth noting in the previous example. The first is the tremendous decrease in the number of photons encountered by the measuring unit, which could be the radiation worker him/herself. From 1000 photons per second measured at 1 meter, only 40 are measured at a distance five times as far from the source. This reduction in air kerma is impressive. In fact, distance is the most effective means of protection against photons because it is unrelated to either the energy or source of photons. So safety can be purely a function of being far enough away.

The second is the assumption that the detector is of fixed size and maintains the same irritation to the source.

The third item worthy of note is that, since the distance units cancel out, any unit distance will do as a reference. It does not matter whether the number of photons per unit time is measured at 1 centimeter or at 1 meter so long as both the reference distance and the distance of interest are measured in the same units.

Finally, this last observation means that, if the number of photons per unit time is known at any distance from the source, the number of photons per unit time that would be measured at any other distance can be calculated using the same simple rule:

$$
\begin{array}{l}
\text{Number of photons per} \\
\text{second at particular} \\
\text{distance from source}
\end{array}
\cdot (\text{distance})^2 =
\begin{array}{l}
\text{Number of photons} \\
\text{per second} \\
\text{at any distance}
\end{array}
\cdot (\text{distance})^2
\qquad (7.5)
$$

Example 7.2

If 2000 photons per second are measured at a distance of 4 meters from a source, how many photons per second will be measured at a distance of 8 meters? Hence:

$$2000 \text{ photons} \cdot 1 \text{ s}^{-1} \cdot (4 \text{ m})^2 = x \cdot (8 \text{ m})^2$$

Rearranging the above equation, one obtains:

$$\frac{2000 \text{ photons} \cdot 1 \text{ s}^{-1} \cdot 16 \text{ m}^2}{64 \text{ m}^2} = x$$

which gives finally:

$$\frac{500 \text{ photons}}{\text{s}} = x$$

In the preceding example, although the distance from the photon source doubled, i.e., changed by a factor of 2 ($2 \times 4 = 8$), the number of photons per second measured decreased by a factor of 2^2, or 4. If the distance had changed by a factor of 3, for example, from, 4 to 12 meters ($3 \times 4 = 12$), the number of photons per second would have decreased by 3^2 or 9. This example emphasizes again that distance is the best protection against photons. In fact, distance is a good means of protection against most kinds of ionizing radiation, since charged particles have a very short range (distance of travel), even in air.

In conclusion, it is a good rule of thumb to put as much distance between a person and a source of ionizing radiation as possible. However, no amount of distance completely eliminates all photons; rather, distance helps reduce the number of photons per unit time encountered to an acceptable, or minuscule, level.

In practical terms, X-ray technologists and physicians should be as far away as possible when the X-ray tube is energized. Nuclear medicine technologist and physicians should stand as far away as possible from a patient injected with radioactive material. A good rule of thumb is to position the patient as quickly and efficiently as possible; then while the image is being taken or the scan is being done, move away, however, always taking patient safety into consideration. The radiation worker in a national laboratory or an industrial setting (e.g., pharmaceutical company, nuclear power plant, or X-ray diffraction application) should examine his/her particular workplace to establish a practical plan incorporating the distance principle.

7.3 Shielding

The third and last type of protection against ionizing radiation consists of placing a barrier, or shield, between the radiation and the rest of the world. This barrier takes many different forms.

To shield from alpha particles, a thin sheet of paper is generally sufficient. For beta particles having those energies commonly encountered, a few millimeters of aluminum or a similar material is sufficient. When dealing with photons, the problem is not so simple. As with distance, no amount of shielding eliminates all photons. In addition, shielding has inherent problems in terms of weight and cost. It is only practical to surround a source of ionizing radiation with so much shielding. For example, when transporting a radionuclide only so much shielding can be used before the package can be moved only with a forklift. Equally, an X-ray machine can be surrounded by just so much shielding before the cost of building the containing wall becomes prohibitive. For these reasons, the effective/equivalent dose limits are used to set standards of irradiation; then distance and shielding are used together to meet these standards.

Note that the discussion of shielding against photons given in this section is purely theoretical. It is intended only to elucidate the concepts described herein. In practice, barrier thickness or shielding, must be calculated as outlined in Chapter 10.

7.3.1 Half-Value Layer

To calculate the amount of shielding required, a concept known as the half-value layer (HVL) is used. This concept applies only in the case of photons.

If a beam of photons consists of only a single energy, the beam is referred to as monoenergetic. A beam of photons may be composed of many energies, generally the case with X-rays. As a result, it is common to speak of quality and quantity of a beam of photons.

Quality refers to the energy of particular photons, and quantity refers to the number of photons of each specific energy. Quality and quantity are described together in a concept known as intensity, defined as the total energy contained in the beam quality × quantity per unit area per unit time. Hence:

$$I \text{ (intensity)} = \frac{E_{\text{Total}}}{\text{Area} \cdot \text{Time}} \tag{7.6}$$

The area referred to is the cross-sectional area of the beam at a particular point in space. Since most beams of photons spread out as they move away from the source, it is important to specify where the intensity is being calculated.

In the following discussion of HVL, a monoenergetic beam of photons has been assumed. The application to a non-monoenergetic (polyenergetic) beam is described in Section 7.3.3.

Consider a beam of photons approaching a barrier, e.g., a simple wall, as shown in Figure 7.2. Some of the photons penetrate to the far side of the wall, and some are attenuated by the wall material. The number of photons attenuated by the wall is a function of both the energy of the photons and the atomic number and density of the material composing the wall. In general, two principles apply:

1. The more energetic the photons, up to a certain energy, the less likely they are to be attenuated.

2. The smaller the atomic number and the less dense the material comprising the barrier, the less likely it is to attenuate the photons.

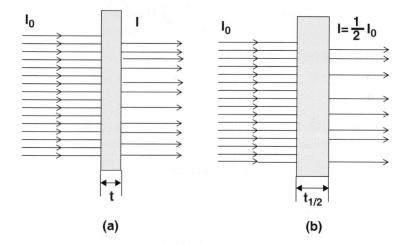

(a)   (b)

Figure 7.2: Half-Value Layer (HVL): (a) A barrier of thickness "t" stops a number of photons consistent with the linear attenuation coefficient μ. (b) Consistent with the linear attenuation coefficient μ of the barrier, a thickness "$t_{1/2}$" is chosen such that half of the photons are attenuated out of the original beam.

For a **narrow** beam of photons, the relationship between the number and energy of photons penetrating the barrier and the material and thickness of the barrier is expressed as follows:

$$N = N_0 e^{-\mu t} \qquad (7.7)$$

N_0 represents the number of photons arriving at the barrier, N represents the number of photons leaving the barrier, t represents thickness of the barrier, and μ is a constant that characterizes the ability of the barrier material to attenuate photons of that particular energy. The constant μ is commonly called the linear attenuation coefficient. It is important to remember that μ depends not only on the barrier material, but also on the energy of photons. The value of μ can be obtained using Eq. (7.7). An example follows.

Example 7.3

A beam of 5000 monoenergetic photons per second is reduced to 3000 photons per second by a barrier consisting of a slab of tin 1.0 centimeter thick. What is the linear attenuation coefficient of the tin slab for these photons?

Using Eq. (7.7) and substituting the quantities given above, $N = 3000$ and $N_0 = 5000$:

$$3000 = 5000\ e^{-\mu t}$$

Rearranging the equation gives:

$$\frac{3000}{5000} = e^{-\mu\,(1\ \mathrm{cm})} = \frac{1}{e^{\mu\,(1\ \mathrm{cm})}}$$

Inverting the preceding expression results in:

$$\frac{5}{3} = e^{\mu\,(1\ \mathrm{cm})}$$

By taking the natural logarithm of both sides (a brief review of logarithms is given in Appendix C) the following result is obtained:

$$\ln \frac{5}{3} = \mu(1\ \mathrm{cm})$$

Using a scientific calculator:

$$0.511 = \mu(1\,\mathrm{cm})$$

and rearranging:

$$0.511\,\mathrm{cm}^{-1} = \mu$$

From this computation it can be seen that μ is measured in units of reciprocal length.

When the thickness t of the barrier is chosen such that the original number of photons N_0 is reduced to one-half of its value, $N = 1/2\ N_0$ then t is written as $t_{1/2}$ and is referred to as the half-value layer (HVL):

$$N = \frac{1}{2}N_0 = N_0\,e^{-\mu\,t_{1/2}} \qquad\qquad (7.8)$$

The HVL, or as it is sometimes called, the half-value thickness (HVT), is a measure of the quality or penetrating power of the beam. The higher the energy of the photons, the thicker the HVL will be.

Dividing both sides of Eq. (7.8) by N_0, a useful relation can be derived between μ and $t_{1/2}$:

$$\frac{1}{2} = e^{-\mu\, t_{1/2}}$$

which gives:

$$\ln 1 - \ln 2 = -\mu t_{1/2}$$

Since:

$$\ln 1 = 0$$

the result is:

$$-\ln 2 = -\mu t_{1/2}$$

Multiplying both sides by minus one gives:

$$\ln 2 = \mu t_{1/2} \tag{7.9}$$

This result leads to two relations. (1) Divide both sides of Eq. (7.9) by μ:

$$\frac{\ln 2}{\mu} = t_{1/2} \tag{7.10}$$

or (2) divide both sides of Eq. (7.9) by $t_{1/2}$:

$$\frac{\ln 2}{t_{1/2}} = \mu \tag{7.11}$$

Eq. (7.10) says that if μ is known for a particular barrier, the thickness of the HVL can be determined. If, on the other hand, the thickness of the HVL is known or determined experimentally, the value of μ can be calculated from Eq. (7.11).

The concept of HVL can now be used to determine how much shielding is needed in any practical application. Dividing both sides of Eq. (7.7) by N_0 yields:

$$\frac{N}{N_0} = e^{-\mu t}$$

Using Eq. (7.11) and substituting the value of μ:

$$\frac{N}{N_0} = e^{-\ln 2 \cdot t / t_{1/2}}$$

Using the properties of logarithms:

$$\frac{N}{N_0} = e^{-\ln 2^{t/t_{1/2}}}$$

one can now write (since $\ln 1 = 0$):

$$\frac{N}{N_0} = \frac{1}{e^{\ln 2^{t/t_{1/2}}}}$$

Finally, because $e^{\ln z} = Z$, one obtains:

$$\frac{N}{N_0} = \frac{1}{2^{t/t_{1/2}}} \qquad (7.12)$$

This equation says that, if the ratio of N to N_0 is known, the number of HVLs needed to produce that ratio can be determined. It is more useful to write Eq. (7.12) in inverted form:

$$\frac{N_0}{N} = 2^{t/t_{1/2}} \qquad (7.13)$$

When the thickness of barrier t is divided by the thickness of a HVL, $t_{1/2}$, the number of HVLs needed is obtained. The original number of photons divided by the number of photons to which the beam is to be reduced then gives two raised to the number of HVLs needed. This result simplifies matters considerably because it is easy to remember the powers of two, which are given are in Table 7.1 up to 2^{10}. To obtain any entry in the 2^n table, just take the number before it and multiply by two. A practical example of the use of Eq. (7.13) follows.

Table 7.1: The powers of 2

n	2^n	$1/2^n$
0	1	1.0
1	2	0.5
2	4	0.25
3	8	0.125
4	16	0.0625
5	32	0.03125
6	64	0.015625
7	128	0.0078125
8	256	0.0039062
9	512	0.0019531
10	1,024	0.0009765

Example 7.4

It is necessary to reduce the source of 5000 monoenergetic photons per second to 1000 photons per second. How many HVL's of a particular substance are needed?

By starting with Eq. (7.13) and substituting numerical values $N_0 = 5000$ and $N = 1000$, the equation becomes:

$$\frac{5000}{1000} = 2^{t/t_{1/2}}$$

Hence:

$$5 = 2^{t/t_{1/2}}$$

From Table 7.1, 5 is between 2 and 3 HVL's. Hence, a little more than 2 HVL's of material is needed to reduce the photon number to the desired value. The actual thickness of HVL is determined by the photon energy and the barrier material. Figure 7.3 gives a graph of HVLs for different materials and photon energies.

In the beginning of this section it was mentioned that distance is often used in conjunction with shielding. Figure 7.4 shows how distance and shielding can be simultaneously used to reduce the photon number to some designated level.

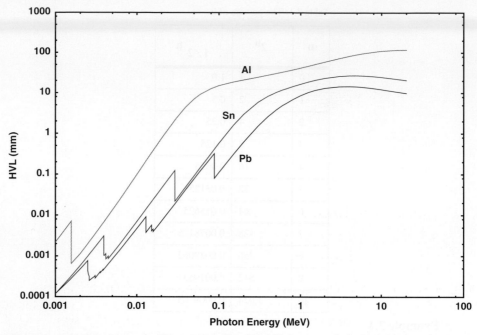

Figure 7.3: Half-Value Layers for Photons 10 keV to 10 MeV: One curve is for lead (Pb), one for tin (Sn), and one for aluminum (Al). (Curves prepared from tabular data published on the United States National Institute for Technology (NIST) web site http://physics.nist.gov/PhysRefData.)

Example 7.5

A monoenergetic source of photons emits 5000 photons per second as measured at a distance of 1.0 meter. A barrier is placed 5.0 meters away. It is necessary to reduce the number of photons per second to 10. How many HVLs thick should the barrier be?

First, the number of photons N_0 arriving at the barrier must be found from Eq. (7.4):

$$5000 \cdot 1\,s^{-1} \cdot (1\text{ m})^2 = x \cdot (5\text{ m})^2$$

which results in:

$$\frac{5000 \text{ m}^2}{25 \text{ m}^2} = x = 200 = N_0$$

Now Eq. (7.13) is used to obtain:

$$\frac{N_0}{N} = \frac{200}{10} = 20 = 2^{\,t/\,t_{1/2}}$$

From Table 7.1 it is evident that a barrier thickness of between 4 and 5 HVL's is needed.

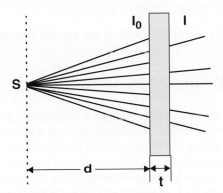

Figure 7.4: Distance and shielding used together to reduce the intensity from the photon source (S)

7.3.2 Tenth-Value Layer

In radiation protection work one also encounters the term Tenth-Value Layer (TVL) or Tenth-Value Thickness (TVT). This term is used to indicate that the shielding used has reduced the radiation source strength to one tenth of its original value. A simple rule of thumb for shielding, particularly for small-volume and weak-strength radioactive sources such as are used in tracer radionuclide work, is that one TVL reduces the radioactive source strength to background level. This rule should be applied only with an educated understanding of its application and use. For sources of very high levels of radiation, exact calculations along the lines just described should be done.

Now the tools developed above are used to derive a relationship between HVL and TVL. Start by writing down a relationship similar to that displayed in Eq. (7.7):

$$N = \frac{1}{10} N_0 = N_0 e^{-\mu t_{1/10}} \qquad (7.14)$$

This leads naturally to the relation:

$$\ln 10 = \mu t_{1/10} \qquad (7.15)$$

Using a scientific calculator, $\ln 10 \approx 3 \cdot \ln 2$ is obtained. Therefore, the above equation may be re-written in the form:

$$\ln 10 = \mu t_{1/10} \approx 3 \cdot \ln 2 = 3\mu t_{1/2} \qquad (7.16)$$

from which it may be concluded that one TVL is approximately equal to three HVL's:

$$1\,TVL \approx 3\,HVL \qquad\qquad\qquad (7.17)$$

This relationship between TVL's and HVL's will be used in the discussion of shielding given in Chapter 10.

7.3.3 Polyenergetic Beams of Photons

Lastly, it is necessary to look at the characteristics of polyenergetic beams of photons. When a beam of photons contains more than one energy, photons that have the least energy are usually stopped first. The weeding out of these lower-energy, "soft", photons changes the quality of the beam; in fact, the beam now contains photons that have a higher average energy. This phenomenon is known as **beam hardening**. Since the lower-energy photons are the first to be stopped, the first HVL of shielding is usually the thinnest, because it is relatively easy to attenuate these photons. As the beam is hardened, more shielding is needed to reduce the number of photons in the beam to one-half. Hence, in a polyenergetic beam of photons, the HVL thicknesses increase as the beam is hardened. Each HVL thickness must be calculated separately. For a close approximation, use the beam intensity in the calculation instead of using the number of photons in the beam and their specific energies. For example, if I_0 is the intensity of the photon beam entering the barrier, then the intensity I of the photon beam leaving the barrier can be calculated by using Eq. (7.8) as follows:

$$I = I_0\, e^{-\mu t} \qquad\qquad\qquad (7.18)$$

The intensity value is naturally based on the average energy of the beam.

Since low energy photons rarely penetrate through patients (the exact low-energy limit depends on the procedure) it is common to use a combination of three metallic filters in front of the anode of an X-ray tube. These filters serve to harden the beam and raise its average energy closer to the peak kilovolt value, thereby preventing the more harmful soft X-rays from reaching the patient. Aluminum alloy filters are often used because they are efficient in removing soft X-rays. This then provides an example of how the intensity of polyenergetic beams is changed by use of beam hardening filters. Such filters are not necessary in industrial applications of X-rays, but may still be used.

7.4 Review Questions

1. What are the practical means of radiation protection?

2. Discuss the use of time as a means of protection.

3. When using the radiation protection principle of distance: a) how does it apply to particulate radiation? b) does one have to consider the energy of the photon beam or the individual photons?

4. What is meant by the inverse square law? To what type of ionizing radiation does it apply?

5. When using the inverse square law, can the reference distance be any distance at which the quantity of radiation is known?

6. What are some practical ways in which the radiation worker can use the protection principle of distance? Answer based on your own environment.

7. What is used to set standards of irradiation for occupationally exposed personnel? What two general protection principles are used together to meet these standards?

8. What is meant by HVL? What is the linear attenuation coefficient? How are HVL and linear attenuation coefficient related?

9. What is meant by the quality and quantity of a photon beam? What concept is used to describe these two parameters together?

10. On what qualities of the photon beam and barrier material do the linear attenuation coefficient and hence the HVL depend?

11. What is meant by the Tenth-Value Layer (TVL)?

12. With respect to a polyenergetic photon beam, what is meant by "beam hardening?" How is the HVL thickness effected by this phenomenon?

7.5 Problems

1. A point source of radiation is measured as emitting 5000 photons per second at a distance of 2 meters from the source. How many photons per second will be measured at a distance of 6 meters from the source?

2. If the number of photons per second from a point source is measured as 10,000 at a distance of 5 meters from a point source, how many photons per second will be measured at a distance of 1 meter from the source? How will the energy of the photons affect this measurement?

3. A beam containing 10,000 monoenergetic photons per second must be reduced by a barrier to a beam containing 1,000 photons per second. How many HVLs will be necessary to accomplish this?

4. The radiation from a monoenergetic point source measured at a distance of 2 meters from the source, consists of 100,000 photons per second. A barrier is placed 4 meters from the source. How many HVLs of material are needed if the photon beam beyond the barrier is to be reduced to 1,000 photons per second?

Chapter 8

PRINCIPLES GOVERNING SPECIFIC DEVICES: GOOD WORKING HABITS

As we saw in the previous chapter, no one method of radiation protection is entirely effective. A radiation worker must be aware of each method and must take best advantage of each. Awareness of and a healthy respect for dangers associated with ionizing radiation are the first steps toward formulating a personal protection plan. Next must be a commitment to the principle, "the best exposure is no exposure". Then comes a firm resolution to adopt and use a personal protection plan. Finally, a thorough knowledge of the principles of radiation protection enables a personal protection plan to become practical reality.

8.1 Responsibility

The ICRP strongly suggests that responsibility for achieving appropriate radiation protection falls not only on employers and equipment manufacturers, but on competent regulating authorities and on radiation workers who are users of products giving rise to irradiation. In some instances, the responsibility may fall on the exposed person as well. In any institution in which radiation protection measures are necessary, a single person is generally designated as the referring local official variously known as the radiation safety officer (RSO), radiation protection officer (RPO), radiation protection supervisor (RPS), health physics officer (HPO), and possibly other titles. This person may be available on a full-time or part-time basis depending on the size of the radiation installation. He/she is responsible for all operations within the institution that involve radiation. The ICRP has recommended that all areas in the institution where a person might receive an annual air kerma of more than three tenths of the occupational effective/equivalent dose limit be designated as a controlled area and that access to that area be limited. This is most easily accomplished by the use of signs. (For the types of sign recommended or required see Chapter 9.) The area itself may be designated by existing structural boundaries such as doors and walls. Within a controlled area the RSO must limit possibility of irradiation primarily by the use of time and/or special protective equipment. In addition, the ICRP recommends that all areas where a person might receive an annual air kerma of more than one tenth of the occupational equivalent dose limit be designated as a supervised area and that access to it also be limited. However, with present standards for medical and industrial facility design, in practice, most areas follow the one-tenth rule, and so there is no difference between controlled and supervised areas and all are designated as controlled.

The ICRP also places importance on the training of radiation workers and emphasizes that this training should stress the radiation protection principles of time (which is considerably enhanced by good technique), distance, and shielding. The training of persons who will plan and implement radiation oncology procedures is crucial. Both the referring physician and radiologist approving the procedure have an obligation to avoid unnecessary procedures to the patient. Toward this end, the ICRP recommends that every medical school curriculum include some training on the basic aspects of radiation risk and efficacy of various radiation procedures. In addition, all institutions where research using radiation is carried out, should have an appropriate expert body to review research programs. Radiation risk to the patient and possibly others versus benefit must be the determining factor in whether or not a radiation procedure is done.

8.2 Facility Design

It is important that diagnostic, oncologic, and industrial radiation facilities be properly designed and maintained. All areas adjacent to radiation producing areas must be checked for compliance with radiation standards, particularly with respect to air kerma.

The effective/equivalent dose limits as defined in Chapter 4 provide the absolute limit, however, the ALARA principle of "As Low As Reasonably Achievable," relevant social and economical considerations being taken into account, is the present acceptable, and in some countries such the United States, the legal standard. Hence, installations should be designed with this principle in mind. In addition, if changes are made, the facilities should be checked once again to make sure that the adjacent areas are as safe as possible. Specific calculations regarding shielding are given in Chapter 10.

8.3 Quality Assurance

To ensure proper functioning of all equipment, quality assurance measures should be instituted and strictly adhered to. The measures outlined here are both for the protection of radiation workers and others (such as patients).

In most diagnostic and many oncologic procedures, the end product is an image captured either in digital form or on a piece of film (more and more, images are only available in digital form). If film is used, a sensitometer or a physical step wedge that exposes film to different density values should be routinely used. The resultant control strip is compared to a standard, preferably through the use of a digital densitometer. In this manner HD [Hurter and Driffield, England 1890] curves demonstrating optical density versus log relative exposure, can be approximated, and measures of film speed, contrast, and base plus fog can be obtained. This also enables verification of the film processor which is the last step in the usual image chain before the viewing of the film. Processor chemistry can change when new batches of chemicals are added as well as when temperature and/or timer-control settings change. Digital thermometers with detachable probes are available. It is recommended that the probes be permanently inserted in the processors and readings taken daily. Additionally, different batches of the same kind of film (blue or green sensitive, etc.) may have slightly different characteristics. Optical density measurements allow these characteristic changes to be detected on a daily basis and to be corrected either by adjusting the processor or by adjusting the examination technique. Figure 8.1 shows a film processor quality assurance instrument set which consists of a sensitometer for producing sensitometer (film) strips; a densitometer which automatically reads the sensitometer strips after development; and a printer on which the densitometer can generate three types of reports. B,

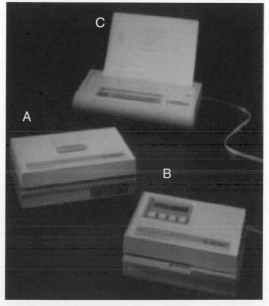

Figure 8.1: The sensitometer **A**, provides dual-color (blue and green), single-sided exposure of sensitometer (film) strips while also allowing adjustment of speed index for variations in film types. The densitometer **B**, which automatically reads each sensitometer strip developed in each film processor to be tested is shown attached to the printer **C**. The densitometer calculates and stores a number of control parameters such as base plus fog, speed index, average gradient, etc., and generates three types of reports: a processor monitoring log, an HD curve with control parameter measurements, and a list of density readings. (Courtesy of X-Rite Inc.)

If a digital image receptor is used, there is usually a designated person or persons who are charged with assuring that the quality of the image is correct. Digital image processing tools, such a gamma correction and various filters are available for post-processing the image. If a particular error is noticed on every image produced by a particular unit, the digital receptor unit should be carefully checked for defects.

In addition, all radiation producing or detecting devices should be acceptance-tested. However, since over time the characteristics of these devices may change, a good quality assurance program will alert the radiation worker to this. For example, a set of instruments such as those described in Chapter 5 might be used to check the quality (kVp) and quantity (mA and time, i.e., mAs) of an X-ray generating unit. Precision test patterns, such as star patterns, may be used to measure the spatial resolution, focal-spot size, and modulation transfer function (MTF) of the X-ray beam generator. Most importantly, the timer circuit should be checked for accuracy, particularly in the case of fluoroscopic, angiographic, and oncologic equipment. The fixed or permanent beam filtering, as well as the result of adding specific amounts of filtering, must be verified. Furthermore, the electrical, mechanical, and toxic hazards that may exist should be checked. If screens are used in a film-screen system, the light output of the screen should be checked periodically to ensure that it is correct. Light devices with known light output that quickly identify screen defects, dirt, and stains are available for this purpose. Cassettes should be cleaned regularly to ensure that no foreign matter is present. If panoramic dental equipment is in use, the alignment of the tube with film shield, which is crucial, must be checked at appropriate intervals. There are some companies now offering comprehensive kits for checking radiography, fluoroscopy, mammography, dental, and CT units. These instrument sets come with dosimeters capable of measuring absorbed dose, absorbed dose rate, kV, time, mA, and mAs - and have full computer control and data capture waveform for kV, FFT, relative absorbed dose rate, and mA. One such instrument is shown is Figure 8.2. In the case of radiation-detection equipment such as gamma cameras, the resolution, linearity, and dead time of the camera should be periodically checked.

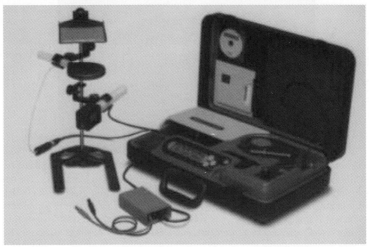

Figure 8.2: A compact and portable system which allows the user to perform all acceptance and quality assurance testing on all diagnostic X-ray generating equipment. (Courtesy of Radcal Corporation.)

For gamma camera measurements, the National Electrical Manufacturers Association (NEMA) has defined a series of tests to be performed on planar, single photon emission computed tomography (SPECT), and positron emission tomography (PET) gamma cameras. These test all aspects of the camera performance and provide a baseline for the ongoing quality assurance program which must be in place for type of unit.

Numerous types of passive mechanical apparati, known as phantoms, are available to check the functional characteristics of every type of radiation emitting or detecting device. Depth-dose phantoms are particularly useful for planning oncologic techniques. The American Association of Physicists in Medicine (AAPM) and the American College Radiology (ACR) have been particularly active in developing phantoms for use with diagnostic equipment, including tomographic techniques such as computed tomography (CT), SPECT, and PET. A representative set of phantoms used in diagnostic radiology in particular, is shown in Figure 8.3.

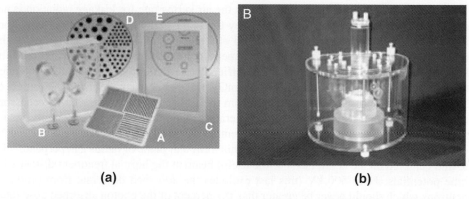

(a)   (b)

Figure 8.3: (a) Numerous phantoms for testing different aspects of planar and SPECT gamma camera: **A** is a resolution phantom, **B** a thyroid phantom, **C** a flood phantom, and **D** and **E** parts of the Jaszczak SPECT phantom. (b) A phantom for testing combined PET-CT units.(Courtesy of Biodex Medical Systems, Inc. (a); Courtesy of Data Spectrum Corporation (b).)

Since image quality (too dark or too light) is one of the major causes of repeat patient examinations, and, hence, a higher absorbed dose for patients, particular attention should be paid to technique. A good quality assurance program alerts both the radiologist, or radiation oncologist, and the radiologic, oncologic, or nuclear medicine technologist to changes in equipment function (including the film processor, if film is used), and alerts supervisory personnel to errors in technique which may be statistically significant (such as an unusual number of repeat examinations) i.e., bad technique (not attributable to equipment failure) which needs to be corrected. This correction can be accomplished by servicing the equipment, adjusting the technique, or instructing all the appropriate personnel in the use of better technique. It is particularly important that fluoroscopic and angiographic equipment be in good working condition. In the case of radiation oncology equipment, a patient can be quite seriously injured by malfunctioning equipment. A good quality assurance program is a necessity. Specific recommendations with respect to performance standards for different types of radiation generating equipment, as well as sources of radioactivity and neutron generation are given below. Particular attention is given is the protection of the radiation worker.

8.4 Specific Recommendations for X-ray Generating Equipment

Whenever X-ray generating equipment is used, particularly for human use, there are certain principles of safe and efficient operation to which all radiation facilities must adhere. These safety measures include leakage of the X-ray tube itself, beam quality, control device function, and protection of both patients and operators.

8.4.1 Leakage, Beam Size Definition, and Beam Filtration

The source tube housing should be such that the air kerma rate for leakage radiation is not exceeded. These values, as recommended by the ICRP and endorsed by most regulatory agencies are: 1 mGy in one hour measured at a distance of 1 m from the focus for the diagnostic equipment; 1 mGy in one hour measured at any position 5 cm from the tube housing or its accessory equipment for superficial radiation oncology equipment intended to operate at voltages within 5 to 50 kV; 10 mGy per hour at one meter from the source, or 300 mGy per hour at any position 5 cm from the surface of the housing or its accessory equipment for radiation oncology X-ray generating equipment with tube potentials greater than 50 kV and less than 500 kV; 0.2 percent of the absorbed dose rate to tissue on the central axis at the treatment distance, but within a circular plane of radius 2 m perpendicular to and centered on the central axis of the useful beam at the normal treatment distance for tube potentials above 500 kV (this last excludes the absorbed dose rate from produced neutrons which should never be greater than 0.1 percent of the photon absorbed dose rate). For equipment not intended to emit radiation, such as high-voltage generators, the air kerma rate must not exceed 0.2 mGy in one hour at a distance of 5 cm from the surface where the leakage measurements are to be averaged over an area not larger than 100 cm^2 at a source distance of 1 m, or 10 cm^2 at a distance of 5 cm from the tube or source housing.

Devices that limit the cross-sectional area of the beam to that of specific clinical interest should always be used. These beam-limiting devices, in combination with the tube housing, must comply with the air kerma requirements for leakage radiation for the particular unit. Any nonadjustable devices must be marked with the appropriate field size and other relevant data. A more extensive discussion of collimation is given in Chapter 10. This is particularly true in the practice of conventional dental radiology where intra-oral films are used. The ICRP recommends that the field diameter never be greater than 7.5 cm and ideally be 6 cm. It is suggested that pointer cones with a fixed field size greater than this be fitted with an appropriate lead annulus in the base. It is also recommended that all field limiting devices have an embedded visible light system that defines the same region that the primary beam will intercept. In addition, it should be possible by use of suitable means provided, to verify that the image receptor is aligned with the beam. In the case of high voltage (150 to 500 kV) radiation oncology equipment, beam-limiting devices should be constructed to reduce, as much as possible, the integral dose (see Glossary) to the patient, transmitting not more than 2% of the useful beam.

For normal diagnostic work, the ICRP recommends a total beam filtration equivalent to 2.5 mm Al with 1.5 mm of this being permanent. When special equipment is used to produce the very soft radiation required for procedures such as mammography, a lower total permanent filtration equivalent to 0.5 mm Al or 0.03 mm Mo is recommended. No beam filtration beyond the permanent 1.5 mm Al filtration is recommended for conventional dental X-ray equipment provided the maximum tube voltage cannot exceed 70 kV. At higher tube voltages, as might be encountered on panoramic or cephalometric equipment, the ICRP recommends that the higher total filtration of 2.5 mm Al be used. Some modern dental equipment has been reduced in size to be used intra-orally. The air kerma at the surface of such tubes has been determined to be as high as 0.5 to 1.0 Gy. Appropriate filtration and extra sensitive film can be used together to produce absorbed doses no higher than 0.05 to 0.1 Gy to the tongue. Used thus, intra-oral tubes may cause lower integral doses to the patient than regular dental tubes and expose the staff less since the entire exam can be accomplished with one tube exposure. Extra shielding in the applicator to limit the radiation field to that of clinical interest can further limit the integral dose to patients. Thus there are radiation protection advantages to this type of equipment.

8.4.2 Control Devices

It is of particular importance that timers of whatever type used function properly. Appropriate levels of reproducibility and accuracy should be maintained, and frequent quality assurance tests should be performed, particularly on mobile and dental equipment. The exposure switch that activates the timer must be such that accidental activation of the X-ray generator is unlikely. In general, repeat exposures after termination by the timing circuit must be possible only by release of the exposure switch to reset the timer. Exceptions to this rule are noted below.

On conventional dental equipment, in addition to the above, the exposure timer should have a maximum range of 5 s and should be capable of reproducing consistently those short exposure times necessitated by high-speed film. With high-energy oncology equipment (150 to 500 kV) a transmission chamber or other device monitoring radiation output constancy should be provided. This will help ensure that the patient has been given the correct amount of radiation.

On fluoroscopic equipment, the ICRP recommends the automatic termination after a preset time, not exceeding 10 minutes. A supplemental audible warning signaling the end of the time period would allow timer resetting in specific clinical situations. However, the total elapsed time during which the beam has been on should be clearly visible to the fluoroscopist. Additionally, the exposure switch must be such that continuous pressure is needed to maintain circuit-closing contact. Control and recording of fluoroscopy time and review of this data is recommended particularly in those institutions in which training programs are in effect.

Each type of X-ray emitting generator should have a control panel that clearly indicates whether or not radiation is being generated. If more than one tube can be activated from a single location, lights displayed on or near each tube should indicate whether or not the tube is presently operating. The ICRP recommends that the control panel also have a way of indicating to the operator, usually with meters, the tube potential, current, and exposure time. If the exposure time is automatic, this should also be indicated. Separate current and time indicators could be replaced by an mAs meter that registers the product of current and time. This is especially important on mammographic units.

On fluoroscopic and photofluorographic units, it is recommended that the values of voltage and current be observable by the operator during the procedure. This is of particular importance when image intensifiers are used which have "automatic brightness control" because the variation of either voltage or current to maintain constant brightness takes place without direct control by the fluoroscopist.

For X-ray oncology units, the control panel should indicate not only the tube voltage and current when variable, but the filtration being used as well. It is recommended by the ICRP that unintentional combinations of tube voltage and filtration be prevented automatically in superficial units (below 150 kV). If practical, preset combinations of tube voltage, current, and filtration should be employed. For machines capable of operating above 50 kV, indicators must be mounted outside the treatment room. Door interlocks of a fail-safe type that interrupt the radiation treatment when the treatment room door is opened, must be provided.

8.4.3 Protection of Patient and Operators

The best protection for patients is good technique. In all types of diagnostic procedures the fastest image receptor of whatever kind should be used. Properly operating equipment is also essential. For the safety of the patient, he/she must always be observable from the control position and provided with a means of communication. Only one patient at a time should be examined or treated in a particular room. No person should be unshielded in a diagnostic procedure room unless his/her presence is essential to conducting the examination or is necessary for training purposes. Essential persons may only remain in a treatment room when a tube voltage of no more than 50 kV is used. In either case, no one should be in the direct beam line. When possible, ICRP and NCRP recommend that all persons be 2 m from where the beam enters the patient to avoid interception of scattered radiation. This applies to mobile equipment and dental installations also. Protective clothing is strongly recommended. Immobilizers should be used as much as possible to restrict patient motion and must always be used for animals. When it is necessary to hold a patient, the recommendations stated in Chapter 7 should be heeded.

Special problems are associated with the use of diagnostic X-ray equipment in surgery and in special procedures such as angiography. Protective devices such as bucky-slot covers, pull-up panels, lead drapes, and/or beaded curtains with interleaving bead layers should be used as much as possible. The beaded curtains can be autoclaved and, hence, sterilized for use in the operating theater. When the hands are inserted through the curtain, the beads part and drape over the physician's wrists and arms, thereby maintaining radiation protection without interfering with hand movements. For the protection of the patient as well as for ease of operation, fluoroscopy should be performed with image intensifiers. Ambient light levels can then be maintained such that visual acuity is not impaired. (When image intensifiers are not used, the eyes should be dark-adapted for at least ten minutes.) By linking together the tube, adjustable beam-limiting device, and fluoroscopic screen or intensifier, it is possible to ensure the beam will not fall outside the image receptor, irrespective of the source-receptor distance. An indication, such as an audible alarm or flashing light, should warn the operator when this system fails. A fluorescent screen must be covered with a protective glass sheet, its lead equivalent being greater for higher maximum tube voltages. Image intensifiers and adjacent fittings should also afford this degree of protection. When the apparatus permits the taking of spot images, the beam should be intercepted by a barrier having a lead equivalent similar to that for a fluorescent screen. This barrier must be designed for the maximum tube voltage at which a procedure can be performed. In addition, the ICRP recommends that while the air kerma rate measured at the patient entrance surface be as low as practical, it should not exceed 5 Gy per minute. If a rate higher than this occurs, there should be clear indication to the fluoroscopist of this higher rate.

Mobile equipment should be used only when it is medically unwise to move the patient. It is important that the operator of mobile equipment ensure that only the patient being examined is exposed to any part of the useful beam. If fluoroscopy is performed with mobile equipment, the use of an image intensifier is highly recommended. The equipment should be such that the radiation beam is fully intercepted by the image receptor area of the image intensifier. It is further recommended that image storage devices be used.

Diagnostic radiology should always be performed in such a way that optimal conditions exist to minimize absorbed dose to the patient, without impairing the diagnostic information. Auxiliary measuring devices to determine focus-skin and/or focus-receptor distance should be provided. When an antiscatter grid is used, particular attention should be paid to align the central axis of the grid properly with the reference axis and, in the case of focused grids, to use a focus-grid distance close to the radius of the grid. To avoid extra absorbed dose to the patient from long exposure times, the ICRP recommends that the aluminum equivalent of the table top, or the front of a vertical cassette holder, must not be more than 1 mm at 100 kV. Carbon fiber material is known to be excellent for transmitting radiation. Hence table tops (particularly for fluoroscopy) and cassettes impregnated with this material are desirable. Ultrasound, if possible, is preferable to ionizing radiation if the patient is known to be pregnant. If pregnancy is possible, but not confirmed, appropriate precautionary measures should be taken. It is strongly recommended that mass-survey photofluorographic equipment be designed and shielded so that all personnel associated with the procedure, and waiting patients, are adequately protected during routine use without need for protective clothing. When a camera is used to photograph an image produced on the fluorescent screen, it is important to achieve maximum sensitivity by using a wide-aperture optical system and the correct image receptor for the type of fluorescent screen. When this equipment is used in association with image intensification, a high efficiency optical and electronic linkage must be maintained. In dental installations the film must be fixed in position or held by the patient, but neither film nor any film holder should be held by the dentist or any member of his/her staff. When conventional, intra-oral films are used, a spacer cone should be used to provide a sufficient focus to skin distance.

8.4.3.1 Protective Clothing

In any technique, diagnostic or oncologic, in which the radiation worker may be exposed, appropriate clothing in the form of an apron, gloves, and sleeves should be worn. The ICRP recommends that these have a lead equivalent of 0.25 mm, and if the workload is especially heavy, that a 0.5 mm lead equivalent be used.

For the protection of the patient, gonadal shields are recommended. However, because of individual differences, these can be a considerable problem. For example, the ICRP has completed a study which shows that the location of the female ovaries, particularly in youths, is variable. Contact shields as well as shaped shields embedded in briefs for men, while not always optimal, are better than nothing. Shadow shields that connect directly to the X-ray tube housing and have a flexible arm are also available for this purpose. These allow angulation of the shield to change its size and position to shield any large or small area of the body and thus can be used for pediatric cases and for shielding of the thyroid as well as the gonads. Cup-shaped lead glass lenses contoured to fit the outer eye orbits snugly and protect the eye lens, even in the lateral position, are available for X-ray examinations in the cranial region. The best protection for both radiation worker and patient is of course: time (good technique), distance, and shielding. Thus, shielding in the form of protective clothing should be used as an extra precautionary measure whenever possible.

8.5 Specific Recommendations for Sealed Sources of Radioactivity

Any radioactive substance sealed or bonded wholly within an inactive material to prevent its dispersion during routine use is known as a sealed source. There are two different methods associated with the use of sealed sources for radiation oncology treatment. When the sealed source is mounted in a housing similar to an X-ray machine, it provides an external collimated beam of photons and this method is known as a collimated sealed source treatment. When the sealed source is attached to or implanted in the patient, the method is referred to as noncollimated sealed source treatment.

8.5.1 Collimated Sealed Sources

When used as a source of collimated photons, a sealed source can be in an unshielded position, known as ON, or in a shielded position, known as OFF. The ICRP recommends that the leakage air kerma rate measured at a distance of one meter from the source not exceed 10 µGy per hour, and that any readily accessible position 5 cm from the surface of the housing have a leakage air kerma rate of not more than 200 µGy in one hour when the source is in the OFF position. The leakage air kerma rate measured at a distance of one meter when the source is ON is recommended to be limited to the greater of 10 mGy per hour or 0.1% of the useful beam air kerma. Performance of the leakage radiation measurements and the use of beam-limiting devices are as described above for X-ray generators.

Appropriate indicators placed at the control panel, at the source housing, and at the entrance to the treatment room should warn that the source is in the ON position. At the end or interruption of treatment, the source should automatically return to the OFF position. Alternate means, usually manual, should be provided to accomplish this in case the automatic system fails. It should also be possible to unload or repair the treatment head without undue air kerma to personnel.

The surface of the housing of the source capsule as well as the beam aperture should be tested periodically for leakage of the source. This can be accomplished as described below for unsealed radioactive sources. The ICRP recommends that this be done at least yearly and that if leakage tests show an activity of more than 2 kBq, the source should be considered leaking, the equipment should then be withdrawn from use and decontaminated, and the source repaired. The decontamination procedures associated with radionuclides are given below in the section on unsealed sources of radioactivity. Fire is a hazard where any radionuclides are involved, so care should be taken to guard against this.

8.5.2 Noncollimated Sealed Sources

The transportation, receipt, logging, and disposal of sealed sources used for radiation oncology follow the rules given in Chapter 9 for any radionuclide. When sealed sources are received, they should be immediately tested for leakage. If leakage is found, the RSO should be notified and proper containment and decontamination procedures as described below for unsealed sources should be followed. Sealed sources should be kept in sufficiently shielded areas as described below, and those which are stored should be tested periodically for leakage. If sealed sources are prepared for use on site from initially unsealed sources, all the precautions given below for unsealed radionuclides should be followed.

Patients who have received diagnostic quantities of unsealed radionuclides are usually cared for in the same manner as other patients. This is so because the activity given is very low and the half-life of the radionuclide is very short. However, patients who have received oncologic quantities of either sealed or unsealed radionuclide sources are possible hazards for the duration of their treatment for two reasons. First, when sealed or unsealed sources emitting high-energy photons are used or when the sources are applied at or near the surface of the patient, the patient is a focus of irradiation for anyone approaching. For this reason, the care of patients who have received radionuclides should be carried out according to a program determined by the RSO. The NCRP recommends that the attending staff wear self-reading monitors, carefully logging the results, and remain in the patient's room for only the time determined by the RSO. Also, visitors to these patients should be instructed to stand approximately 2 m from the patient, with a limited length of stay. In addition, patients receiving unsealed sources, such as for treatment of thyroid disease, are a source of contamination. For this reason, any bodily excrements (vomit, sweat, urine, etc.) contain a high percentage of radioactivity. Thus the sheets, eating utensils, sanitary facilities, etc., of these patients must be dealt with according to a program prescribed by the RSO. For patients who have received sealed radioactive sources, on the other hand, these precautions do not normally apply. However, sealed sources are the foci of contamination if leakage occurs or if the source dislodges. If leakage or a dislodged source is suspected, nothing such as bed sheets, medical dressings, etc., pertaining to the patient should be discarded until the RSO and the attending physician have determined that it is safe to do so. Radium is undesirable as a radionuclide for a sealed source because its radioactive decay product radon, is a gas that tends to diffuse through the sealed container. Radium implants and applicators which were previously very popular, are gradually being replaced by other radionuclides.

8.6 Specific Recommendations for Particle Accelerators

Particle accelerators are generally used to produce very high energy X-ray photons (megavolts and higher) and electrons. The major concern when using accelerators is to minimize secondary radiation and, hence, the choice and arrangement of absorbing material is of special importance. The absorbed dose inside and outside the treatment area should be kept as low as practicable. The ICRP recommends that the adjustable beam-limiting devices be constructed so that the leakage radiation imparts less energy to the patient than is imparted by a treatment area of 10 cm^2. For the patient's safety, two independent dose-monitoring systems are recommended. Failure or malfunction of one system should not influence the other. They should be designed so that if one system fails to terminate the accelerator at the end of the treatment period, the other will do so after an additional 0.4 Gy. Since accelerators are complex instruments, there should be a control panel that displays the parameters essential for correct performance of a radiation treatment. Mistakes in the selection of types and energy of radiation wedge filters, scattering foils, etc., should be prevented by a system of interlocks. The control panel should also show the preselected number of monitor units and the accumulating number. With high-energy accelerators working above 10 MeV, a significant number of neutrons are produced by photons interacting directly with the atomic nucleus (photodisintegration). No single satisfactory method exists for measuring the contribution of neutrons to the patient-absorbed dose, or for measuring the equivalent dose rate outside the treatment room, so more than one method should be used. It has been known that the equivalent dose rate of neutrons outside the treatment room is significant, so care should be taken to prevent this both in the facility design stage and during radiation protection surveys. Through nuclear reactions (see Chapter 1) neutrons may cause materials in the treatment room to become radioactive (this process is known as neutron activation). Proper precautions should therefore be taken when doing maintenance in the treatment area.

8.7 Specific Recommendations for Neutron Generators

Neutrons are used in medicine for both diagnostic and oncologic applications. Since neutrons deliver a higher equivalent dose to the patient, their use should be limited to those diagnostic applications in which other methods are not applicable. The measurement of neutrons, both for dosimetry and protection purposes, is complicated. Neutron shielding, particularly for the protection of the radiation worker, is more difficult. Hence, special training in the use of neutrons is indicated. The ICRP recommends that the tissue kerma in air due to leakage radiation from neutron generators should be less than 1% of that in the primary beam at the same distance.

As mentioned above, one of the primary ways in which neutrons interact with matter composed of low atomic number elements is through nuclear reactions. Since this interaction often produces radionuclides which emit photons or other ionizing particles, the radiation worker must be aware that the patient and objects in the room with the patient may be radioactive. The ICRP recommends that objects frequently used there, be left in the treatment room and that use of metals containing manganese and glass (containing sodium) be avoided. The radioactivity induced in patients is generally short-lived and no special precautions regarding care of the patient are recommended.

8.8 Specific Recommendations for Unsealed Sources of Radioactivity

Wherever unsealed sources of radionuclides are present, the principles of protection take on a twofold aspect: protection from external air kerma and protection against internal air kerma. A radionuclide present in a working area is outside the body and thus constitutes a potential external hazard. This radionuclide becomes an internal hazard when it is ingested, absorbed through the skin, or inhaled. Thus when working with unsealed sources of radioactivity it is particularly important that the radiation worker be aware of the dangers that are present and in particular test for and guard against unwanted contamination of the both the work place and the persons present in the workplace or under going examination or treatment.

8.8.1 Common External Sources of Irradiation

The most common source of external irradiation is photons, and both radionuclides and X-ray machines are possible sources of photons. Radionuclides can be a source of beta and alpha particles as well.

Photons with energies greater than 100 keV are highly penetrating and can irradiate (expose) the entire body fairly uniformly. Photons that have smaller energies have less penetrating power and give a more superficial air kerma, or absorbed dose, to the body. The air kerma from low-energy photons is more localized in the sense that only a limited number of layers of skin or of skin plus other tissue may be irradiated.

On the other hand, the external danger from alpha-emitting radionuclides is small. Even alpha particles with energies of up to 5 MeV would not be able to penetrate the outermost (dead) layers of the skin.

An external source of beta particles is a cause for more or less concern, depending upon the energy of the particles. Beta particles are very light when compared to alpha particles, but they do have charge. As a result of their light mass, they penetrate farther into any material than alpha particles do before losing their energy through ionization. Beta particles themselves can have a range in tissue up to 1 cm, but they rarely penetrate any farther into the human body than the outer layer of the skin (see Table 1.2). The more pressing danger from beta particles is associated with the production of photons.

As stated in Chapter 1, when high-energy electrons (a few keV and up) pass through matter, they are attracted, because of their negative charge, by the nuclei of the atoms of which the material is composed. This attraction deflects the electron from its path, and as a result, the electron radiates some energy in the form of a photon. This process results in a slowing down of the electron, or a braking action, commonly known as bremsstrahlung. Heavy nuclei found in shielding materials such as lead, have a greater probability of causing bremsstrahlung. Consequently, the container holding a radionuclide that is a beta emitter may be itself a source of photons. For this reason, it is common to store beta emitting sources in aluminum containers and then to place this container inside a lead one.

8.8.2 Protection from External Sources of Irradiation

To protect oneself from the external sources just described, the principles of time, distance, and shielding must be used. Unsealed radionuclides, particularly photon emitters, should be carefully stored in a lead container or behind lead bricks. It is normal to store small quantities of radionuclides, in either liquid or pill form, in individual vials, the vials then being placed inside small lead containers. These containers should in turn be placed in a lead-lined storage area, either into a heavily leaded container (sometimes containing drawers) known as a "pig" or behind lead bricks.

When a vial is needed to administer a determined amount of radionuclide to a patient, it is taken out of the lead container or from behind the lead bricks and the quantity required quickly withdrawn (or the pill removed). If the radionuclide is in liquid form, the amount of radioactivity to be administered (commonly called a "dose") is drawn up in a syringe. This syringe should have a lead glass shield around it, and it should be placed as quickly as possible into a lead container until the patient is injected. For the actual extraction of the radioactivity, the facility should have a tabletop lead shield with a lead glass viewing area on top. This type of shield protects the whole body of the radiation worker, even the sensitive lens of the eye, while still allowing full visualization of the working area. If the hands are used directly, lead-lined gloves should be worn. If these gloves are too awkward, plastic gloves should be used for purposes of contamination isolation (as discussed later in this chapter). It is ideal to use a pair of tongs so that even the hands are not irradiated. This last is an example of using the principles of shielding and distance together. Figure 8.4 shows a radiation worker using tongs inside a chemical fume hood. In this case, radiochemist Sharon Stone–Elander, Ph.D. uses mirrors to watch what she is doing and, hence, is able to stand back from the opening in the lead brick shield.

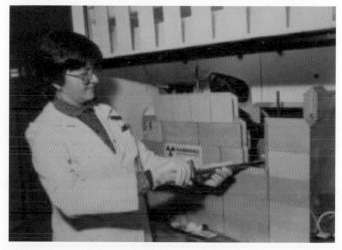

Figure 8.4: Illustration showing tongs, interlocking lead brick shield and mirrors to exemplify the use of the protection principles of shielding and distance when working with radionuclides. (Photograph by Fia Hedér.)

In laboratories or clinics where radionuclides are used, they should always be stored in a separate area. Whenever radionuclides are used, they should be shielded as well as practicable; the time of possible irradiation should be reduced as much as possible by employing efficient working techniques; and tongs should be used along with other methods to increase the distance from the radionuclide.

A patient who has been given a photon-emitting radionuclide is also a source of external air kerma. A radiation worker should be aware of this aspect and should position the patient for the examination as quickly as possible, then move away. Hovering over the patient unnecessarily only tends to increase the possible irradiation of the technologist. The patient should be cared for as meticulously as is appropriate, but the principles of time, distance, and shielding should also be applied. Some radiation workers may wish to wear a lead or tin apron. In some institutions lead or tin screens are placed near the patient while the study is being conducted.

The ability of any particular material to attenuate photons depends on the linear attenuation coefficient for that material divided by its density. It is instructive to study this for lead. From Figure 8.5 it is apparent that a sharp drop in the ability of lead to attenuate or absorb photons occurs in the energy region of about 3 MeV. The reason for this drop is that in lead the probability of a photoelectric or a Compton interaction decreases rapidly as the photon energy approaches 3 MeV, and at the same time the probability for a pair-production interaction has not increased sufficiently. This characteristic of lead should be kept in mind when shielding is designed for photon sources, such as high-energy radiation oncology machines or high-energy radionuclides.)

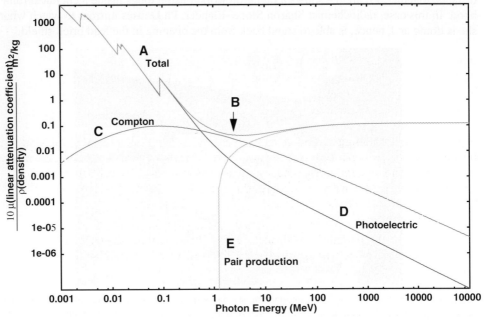

Figure 8.5: Graph showing the decrease in the ability of lead to attenuate photons whose energy is about 3 MeV. Note the decrease in the total attenuation coefficient **A** at **B**. **C** is the Compton scattering attenuation coefficient; **D** is the photoelectric attenuation coefficient; **E** is the pair production attenuation coefficient. All linear attenuation coefficients are divided by the density of lead.(Curves prepared from tabular data published on the United States National Institute for Technology (NIST) website http://physics.nist.gov/PhysRefData.)

8.8.3 Common Internal Sources of Irradiation

Alpha particles, because of their large mass and double positive charge, have a very short range both in air and in tissue. The range of alpha particles, even with energies as high as

4 to 9 MeV, is only a few hundredths of a millimeter in tissue and a few centimeters in air. Alpha particles are highly ionizing, and since their range is short, all this energy is deposited in a very small region near where the alpha particle is emitted. Further, as was discussed in Chapter 3, the radiation weighting factor (i.e., the measure of biological damage) of alpha particles for tissue is considerably greater (about 20 times) than that of photons or beta particles.

In short, while alpha particles pose no serious external threat, when deposited throughout a vital organ, they can cause considerable damage. What is more, many alpha-emitting radionuclides concentrate in tissues like cortical and trabecular bone, usually on the surface. Bone has an organic component and a mineral matrix. The turnover rate is different for each. Thus, depending on where the alpha emitter concentrates, the activity could decrease only as fast as the physical half-life of the radionuclide, since the physiological bone tissue turnover rate in adults is extremely low. As a result, a radionuclide that is an alpha emitter can still cause damage long after it has been admitted to the body. For these reasons, long-lived alpha emitters are poor agents for diagnostic purposes. However, they are a good agent for radioimmuniotherapy, if they can be delivered to the tumor itself and not to any other tissue or organ.

Beta particles, because they are much lighter than alpha particles and have a single charge, have a longer range in both air and tissue and are 20 times less destructive; however, beta particles are directly ionizing particles, and rarely is their range long enough for them to leave the organ in which they are produced. As a result, they deposit all or most of their energy in that organ. Should this organ be a vital one and should the quantity of beta emitters in the organ be large, serious consequences could result. Beta particles are also a good candidate for radioimmuniotherapy applications and are frequently used for patients who have thyroid disease or thyroid cancer locally or metastatically because of the ability of thyroid tissue to absorb (in this case, radioactive) iodine.

One example of a possibly dangerous beta emitter is strontium-90 (^{90}Sr), which has a radioactive (physical) half-life of 27.4 years. This radionuclide is produced in fission reactions, and large quantities of it have been observed in the radioactive fallout following nuclear weapons testing. The nuclide that results from the beta decay of ^{90}Sr, yttrium-90 (^{90}Y), is also a beta emitter. Strontium is a bone seeker and this factor, coupled with its long physical half-life, makes it an undesirable internal radionuclide. So far, the total absorbed dose contamination to the entire world population with ^{90}Sr from nuclear fallout (weapon testing as well as accidental release from power plants and nuclear research facilities) has been less than 100 Gy.

A second example is ^{131}I. This radionuclide is commonly used in nuclear medicine because iodine is a thyroid seeker. It is, however, both a beta and a gamma emitter. The radioactive half-life of ^{131}I is approximately 8 days. This relatively short half-life, together with the high physiological turnover rate of iodine in the thyroid gland, makes ^{131}I less of a hazard than, say ^{90}Sr. However, ^{90}Sr and ^{131}I are major components of the gaseous effluents from fission reactors, and ^{131}I is frequently used in medicine for diagnosis of thyroid disease and always used when radiation treatment of thyroid disease is prescribed.

Therefore, most radiation workers have a high probability of encountering one or both of these radionuclides.

Photons, because of their relatively low ability to ionize, deposit much less energy per unit path length in tissue than do either alpha or beta particles. For this reason, radionuclides that are pure photon emitters represent much less of an internal hazard than do either alpha or beta emitters. Radionuclides that emit alpha and/or beta particles generally emit photons as well. In that case, most of the absorbed dose is due to the particulate radiation, and not much is attributable to the photons.

8.8.4 Common Methods of Acquiring Internal Irradiation Sources

The three common methods of unintentional internal acquisition of radionuclides are inhalation, ingestion, or absorption. When radionuclides are intentionally administered internally, they generally enter the body by injection, inhalation, or ingestion. Here, only the unintentional aspects of entry to the body are discussed. Table 8.1 gives the amount of radioactivity allowed per cubic centimeter (cc) of air and of water for some selected radionuclides. These Maximum Permissible Concentrations (MPC) are calculated from the effective/equivalent dose limits, based on the body burden that each would give. Thus, for occupationally exposed personnel the concentrations are based on the amount of radionuclide an average person breathing and drinking normally would acquire in a working day of eight hours and thus a working week of forty hours. These numbers, as well as the Maximum Permissible Body Burdens (MPBB) stated in Table 8.2 are derived from the accumulated absorbed dose resulting from having this body burden steadily over a working life-time of fifty years. The implication of a higher body burden isolated in time (i.e., not a continuous state of body burden) must be evaluated separately (see Chapter 11). For the general public (as opposed to the radiation worker) the figures given in Table 8.2 should be divided by ten for those who could be irradiated infrequently and by fifty for those who could be irradiated continuously. The method of calculating both the absorbed dose from internally deposited radionuclides and, hence, the MPC and MPBB, was pioneered by the International Commission on Radiological Protection in ICRP Publication 2, undated in ICRP Publications 30, 60, and 61. In Publication 30, the Commission defines the Annual Limit of Intake (ALI) and recommends that it be used for defining limits. The MPC can be derived from the ALI.

8.8.4.1 Inhalation

Inhalation of airborne materials is a serious internal hazard. The absorption, retention, and elimination of material taken into the lungs depend on such things as the size of the particle inhaled, its solubility, and the rate of respiration of the individual inhaling it. Persons should particularly guard against inhaling radionuclides in the form of aerosols, fine powders, or volatile liquids. Iodine is a particularly volatile liquid, and radioactive isotopes of iodine in liquid form are frequently used in chemical laboratory tests (*in vitro*) as well as in diagnostic tests and radiation oncologic treatments (*in vivo*) where they are administered directly to patients.

Table 8.1: Maximum Permissible Concentrations of Radionuclides in Air and Water[a]

Unrestricted Areas				
Radionuclide	Water[b]		Air[b]	
	$Bq \cdot cc^{-1}$	$(pCi \cdot cc^{-1})$	$Bq \cdot cc^{-1}$	$(pCi \cdot cc^{-1})$
^{14}C	2.95	(80.0)	$3.70 \cdot 10^{-3}$	(0.1)
^{3}H	111	$(3.0 \cdot 10^{3})$	$7.40 \cdot 10^{-3}$	(0.2)
^{131}I	$1.11 \cdot 10^{-4}$	(0.3)	$3.70 \cdot 10^{-6}$	$(1.0 \cdot 10^{-4)}$
^{125}I	$7.40 \cdot 10^{-3}$	(0.2)	$2.96 \cdot 10^{-6}$	$(8.0 \cdot 10^{-5})$
^{32}P	0.74	(20.0)	$7.40 \cdot 10^{-5}$	$(2.0 \cdot 10^{-3)}$
^{51}Cr	7.4	(200.0)	$2.96 \cdot 10^{-3}$	$(8.0 \cdot 10^{-2)}$
^{133}Xe			$1.11 \cdot 10^{-3}$	(0.3)
Restricted Areas (40-hour week)				
^{14}C	$7.40 \cdot 10^{-2}$	$(2 \cdot 104)$	0.148	(1.0)
^{3}H	$3.70 \cdot 10^{3}$	$(1 \cdot 10^{5})$	0.185	(5.0)
^{131}I	2.22	(60)	$3.33 \cdot 10^{-4}$	(0.009)
^{125}I	1.48	(40)	$1.85 \cdot 10^{-4}$	(0.005)
^{32}P	18.5	$(5 \cdot 10^{2})$	$2.59 \cdot 10^{-3}$	(0.07)
^{51}Cr	$1.85 \cdot 10^{-3}$	$(5 \cdot 10^{4})$	$7.40 \cdot 10^{-2}$	(2.0)
^{133}Xe			0.37	(10)

[a] Adapted from the U.S. Code of Federal Regulations, Title 10, Part 20, as of November 2006.
[b] cc = cubic centimeter = cm^{3}.

Consequently, whenever volatile liquids are stored or used, a chemical fume hood should be used. The inner surfaces of this hood should be made of a nonporous material, such as stainless steel, or be covered with a paint that is easy to strip off so as to permit easy decontamination. A fume hood that vents radioactive materials must have its output strictly

Table 8.2: Maximum Permissible Body Burdens for Occupationally Exposed Personnel[a]

Radionuclide	Total Body Burden		Critical Organ Burden		Critical Organ
	MBq	(mCi)	MBq	(mCi)	
^{14}C	14.8	(0.4)	11.1	(0.3)	Fat
^{3}H	74.0	(2.0)	37.0	(1.0)	Body tissue
^{131}I	1.84	(0.05)	$2.59 \cdot 10^{-2}$	$(0.7 \cdot 10^{-3})$	Thyroid
^{32}P	1.11	(0.03)	0.222	(0.006)	Bone
^{51}Cr	2.95	(0.8)	37.0	(1.0)	lung

[a] Reproduced with permission of the U.S. Department of Health, Education and Welfare, Public Health Service, Food and Drug Administration, Bureau of Radiological Health, from the Radiological Health Handbook, rev. ed., 1970. Taken from National Bureau of Standards Handbook 69, U.S. Government Printing Office, Washington, DC, 1963.

Also see ICRP, Publication 17, Protection of Patients in Radionuclide Investigations. Amsterdam, The Netherlands, Elsevier Inc., 1971 and NCRP Report 22, 1959, and Addendum, 1963, Washington, DC, National Council on Radiation Protection and Measurements.

regulated. The amount of radioactivity allowed per cubic centimeter is given for some selected radionuclides in Table 8.1. Rarely are such quantities a problem, except in the case of accidents.

As an alternative to a fume hood, a totally enclosed chamber, known as a glove box, may be used. The advantages of a glove box are not only that it confines any contamination within a totally enclosed space, but also that it uses a very small air supply to prevent the escape of activity into the working atmosphere when compared to the generous airflow necessary in the usual hood system. Among the disadvantages are the extra time and labor involved and the difficulties of working inside the box, as well as those associated with removing the radionuclide from the glove box. On the other hand, glove boxes are indispensable for working with dry powders and are highly recommended for work with alpha emitters and low-energy beta emitters such as tritium.

It is likewise desirable that all equipment needed for the handling of radioactive material be segregated and used only for this work. One method of accomplishing this objective is to tag as "radioactive materials" the various items used. It is particularly important to tag small items such as glassware and handling equipment such as tongs, which can easily transfer contamination from fume hoods to the open laboratory.

8.8.4.2 Ingestion

Ingested material can enter the body by absorption from the gastrointestinal (GI) tract. The chemical and physical form of the ingested material determines the percent that is absorbed by the bloodstream. A large proportion of insoluble ingested material is rapidly excreted in normal body wastes. A significant enough internal hazard may be caused simply by the irradiation of the gastrointestinal tract itself.

It is a good radiation-protection practice never to take unessential personal items into the active area. All food, drink, smoking materials, and cosmetics should remain outside the working area, including the scanning rooms; and nothing should be placed in the mouth while working in an active area. This practice is essential when potentially contaminated equipment is used. Such equipment should never be handled with bare hands. In particular, pipettes should never be operated by mouth suction. Any type of glass making in active areas or on contaminated equipment should be done with special techniques that avoid forming the glass by blowing with the mouth.

8.8.4.3 Absorption

Absorption through unbroken skin or through abrasions, cuts, and punctures is still another way that unwanted radioactive material may enter the bloodstream. To prevent this type of entry, all personnel working with radionuclides should follow proper procedures and wear protective clothing, particularly gloves, to prevent contact between radionuclides and the skin. Patients who have received an internal dose of a radionuclide often excrete some of this material in their sweat. It is, therefore, important that the radiation worker wear gloves not only when handling the radionuclides, but also when attending to radioactive patients, as various pharmaceutical forms such as pertechnetate go readily to the sweat glands.

In this same vein, all personnel working in radiation areas must wear proper protective clothing. The clothing change considered necessary depends on the level of activity and the types of procedure performed. In ordinary tracer laboratories, a standard laboratory coat is sufficient. In intermediate-level laboratories, such as a radiation oncology laboratory, coats and a change of shoes may be compulsory; or a complete change of clothing plus a shoe change may be required.

8.8.5 Protection from Internal Sources of Irradiation

The radiation worker must take precautions against the deposition of radionuclides within the body. Internal radioactive material produces continuous irradiation until it physically decays or is eliminated through normal metabolic processes. The amount of internal irradiation is a function of the decay mode (type and energy of radiation emitted) and amount of radionuclide inside the body. In addition, the internal hazard depends upon the size, shape, and biologic importance of the absorbing organ(s) and/or tissue(s) within the body, and the length of time that the radionuclide remains there. In Chapter 11, the hazard in terms of absorbed dose received from internally deposited radionuclides are evaluated.

Internal air kerma control is exercised by preventing the entry of radioactive material into the body. The control of internal irradiation, therefore, consists in preventing radionuclides from entering the body; hence, internal irradiation control is essentially a problem of contamination control. Two of the fundamental principles of such control are containment and cleanliness.

Containment is that form of radioactive contamination control which seeks to restrict radioactive materials to specified areas. The choice of operating techniques for handling radioactive materials is an important part of achieving containment goals. For instance, when working with radionuclides in liquid form, the working surface should be covered by either plastic trays or a layer of disposable absorbent material to soak up spills. Materials used extensively include blotting paper and diaper paper, which is heavy absorbent paper backed with an impervious material such as oiled paper or plastic film. Diaper paper is preferable because it prevents liquids from ever reaching the bench top.

Additionally, when introducing a new procedure, practice it several times with inactive materials before undertaking any manipulations involving radionuclides. Unexpected difficulties are uncovered in this manner, and equipment weaknesses detected. Thus procedural modifications can be made as necessary. What is more, single containers should never be used. Instead use suitable trays or double containers that are capable of holding the primary containers entirely.

Cleanliness and good housekeeping are the most effective supplements to proper technique. These practices not only minimize the spread of contamination, but also prevent the buildup of significant levels of contamination. Cleanliness is not just associated with visible good housekeeping; it also assumes a regular monitoring of the laboratory and of the personnel followed by prompt decontamination when necessary. Working surfaces and floors should be checked regularly. Although local regulations may vary, daily surface checks are frequently mandated. Portable survey instruments are sufficient to monitor bench surfaces and floors where no sources of sufficient strength to cause a high room background are stored. The survey instrument chosen should be one that has the proper response for the type and energy of radiation being used. End-window G-M tubes are the most commonly used instruments for this type of monitoring. Additionally, more elaborate surface checks known as "wipe" tests are sometimes locally mandated weekly (the US NRC has phased out this requirement). These tests consist of using an alcohol pad or other suitably moistened small cotton or other swab, rubbing the swab over selected areas, and then evaluating the radioactivity contained on the swab by counting it in a well type scintillation counter or another device with equivalent sensitivity. It is important to know the efficiency of the counter used for wipe tests as the exact amount of radioactivity detected must be recorded.

If a sufficiently high background exists, or if low-energy beta emitters like tritium (^{3}H) or low-energy photon emitters like ^{125}I are involved, the detection of traces of contamination with a survey instrument is difficult. In this case, surfaces are wiped with small pieces of filter paper or other suitable material. These wipes are them removed and counted with an appropriate counter; ideally, a well counter or a liquid scintillation counter is used. In either case, the counter selected must have a detector that is sensitive to the type and energy of the radiation involved.

It is good practice to have everyone working with radionuclides wash and monitor his/her hands before leaving the area to eat. Radiation workers should be checked regularly by some sort of whole body counting method to determine their internal radiation body burden. Keep in mind that the allowable internal body burden is different for different radionuclides. The maximum permissible body burdens for some radionuclides are summarized in Table 8.2.

8.8.6 Specific Contamination Control Devices

Many products are commercially available to aid the radiation worker in guarding against irradiation or contamination by external radionuclides. Additionally, there are many products available to aid in decontamination.

8.8.6.1 Shields

Syringe shields of various sizes made of lead with a lead-glass window can be used for extracting a given quantity of radioactive material from a vial or other similar container. The syringe shield should appropriate for the energy level of the radionuclide, i.e., different shields are required for positron emitters such as ^{18}F used in PET procedures, than are required in nuclear medicine facilities which do not use positron emitters. The syringe can then be placed in a lead-lined container to transport it to the place where it is to be used. Shields are also available for holding a vial of radioactive material. This container can be placed around the vial before the radioactive material is inserted into the vial. It then stays in place when the vial is calibrated, transported, and used for extraction of a determined amount of radioactivity. These containers usually have a lead-glass viewing window.

For storing containers of radioactive material, there exist lead-lined storage compartments. Some of these are small units with a handle for easy carrying and a lead-glass window area so that the contents of the storage module can be determined without opening it. Large stationary storage modules also exist. These are often in the form of a bench top, with or without a sink unit, with individual lead-lined drawers extending from the bench top to the floor. These drawers can be easily labeled to identify the contents. Such units can also be used to store radioactive waste material until the activity has been reduced to that of background radiation. Frequently, lead bricks are used on a counter or table top to form a storage area for individually shielded vials. These bricks are available in standard design or in interlocking design. The interlocking bricks, shown previously in Figure 8.4, eliminate the danger of radiation leakage where the bricks are joined and also have a smaller chance of toppling. Figure 8.6 shows bricks surrounding a titration apparatus where again the radiation worker uses mirrors to operate the apparatus in order to be better protected. Either type of brick can have embedded into it lead glass windows of various sizes for viewing behind the bricks as well as different size enclosures for insertion of remote handling tools through the shield. Figure 8.7a shows a radiation worker using such remote handling tools and lead-glass windows to manipulate very high activity radionuclides. In this case, the shield is a solid sliding lead door enclosing a chemical fume hood. Figure 8.7b shows the lead shielding removed (when there is no radioactivity present) for adjustment of the apparatus. Simple lead-glass bench top shields are available in addition to solid lead body shields with a lead-glass top. Portable floor mounted lead and lead glass shields for placement near a radioactive patients are available. These shields come in varying ratios of lead to lead glass. Furthermore, wheel chairs with lead-lined backs of various sizes can be purchased.

Figure 8.6: A titration apparatus shielded by lead bricks for maximal radiation protection. Mirrors are used so that the radiation worker can advantageously use the shielding provided. (Photograph by Fia Hedér.)

Additionally, lead-lined, often spill-proof radioactive waste containers are available. These usually feature an inner liner that can be removed for emptying and for decontamination if necessary. An additional plastic liner for easy transfer of waste is used. One good feature to note in a waste container is that it is spill-proof, i.e., if the container is accidentally tipped over, it does not open. Large waste storage drums are usually kept in lead-lined modules of suitable size. These may even have bench tops for working area.

For storing perishable radioactive materials, a variety of lead-lined refrigerators are available. Note that lunch should not be stored along with the radionuclides! Lead foil which is easily cut with a kitchen scissor and molded into various shapes is another very useful accessory. It can be used for many temporary storage jobs.

8.8.6.2 Remote Handling Equipment

Hand-held tongs of various lengths and ending in various jaw assemblies are available. It is a good idea to practice with this type of equipment to become facile with its use.Many types of patient holders are also available. This enables the technologist to reliably position the patient and move away. One versatile type of patient positioner consists of a thin-walled, polyurethane mattress filled loosely with expanded polystyrene beads. After

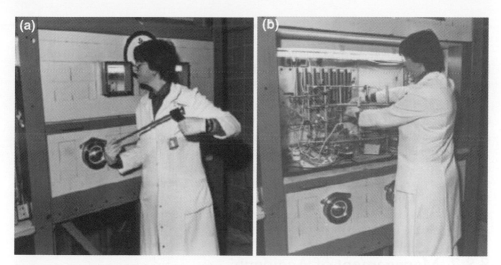

Figure 8.7: Illustration of a completely shielded working area. (a) When the shield is closed, the radiation worker inserts tongs through a "safe" slot and views her work through lead glass windows. (b) The shield opens (when there is no radioactivity present) to allow adjustment of the apparatus. (Photograph by Fia Hedér.)

the patient is properly positioned a vacuum pump evacuates the air and the mattress solidifies into a rock-hard mold that conforms to the contours of the body and permits no movement. Opening the air vent restores the mattress to a pliable state. Half-size mattresses are useful for pediatric patients.

8.8.6.3 Decontamination Equipment

When using liquid reactive material, contamination in the form of spills, even just a drop, is a problem. It is good to work with the radioactive material on absorbent paper as well as to have a protective tray of easily decontaminated material beneath the paper. Both the absorbent paper and the trap help to contain any spill. The paper (and the tray if necessary) can be stored with the radioactive waste for future disposal. The tray can be decontaminated using detergents especially made for surfaces. If the activity of the tray is very high, it can be stored until a survey instrument or wipe test shows its activity has dropped to background, or it can be disposed of properly. Similarly, metal planchets used as sample holders can be disposed of or allowed to decay.

For cleaning up spills on surfaces such as counter tops or floors, some detergents exist which will solubilize and remove reactive contamination. Problem substances such as tars, waxes, resins, dried blood, and greases can be removed with proper agents from surfaces composed of glass, metal, plastic, or rubber. Survey instruments and/or wipe tests should be performed to determine that the surface has been reduced to background radiation. Otherwise the area should be marked as unsafe. Special tape and tags are available for such designation.

For cleaning up spills on the skin, special foam detergent cleaners are available which lift contaminants from pores and skin surfaces and hold them in solution until rinsed away. These detergents work well in cold to tepid water (remember hot water will bring the blood vessels closer to the skin surfaces thus increasing the probability of absorption through the skin) and have no abrasives or skin irritants.

Protective clothing is always recommended when working with radioactive liquids. Gloves, usually rubber and disposable are the most important item. If an area is to be decontaminated from a spill or there is danger of liquid splashing, protective coveralls, perhaps even with a hood, and shoe covers or boots to go over the shoes should be worn. If there is a possibility that the area being surveyed or decontaminated, contains airborne radioactivity, gas masks of several varieties are available. The proper one should be chosen. For example, for radioactive gas contamination, a completely closed system with its own air supply may be appropriate. If particulate airborne matter is known or suspected to be present, suitable filters should be worn with the mask.

8.8.7 Management of Accidents

In the event of an accident involving the contamination of an area, personnel protection and the immediate confinement of the contamination are of primary importance. A standardized approach should be well-known and followed by all personnel in the laboratory or clinic. Emergency procedures should be posted in each area along with the person, usually the RSO, to be contacted in case of an accident involving radiation. A typical plan follows.

8.8.7.1 Confinement

The spread of contamination should be prevented by:

1. closing doors and windows.

2. turning off fans, air conditioners, and other ventilation if possible.

3. closing ventilation ducts if possible.

4. vacating the room, but not the area, so as not to spread the contamination; in case of a liquid spill, leaving shoes and other garments at the door.

5. locking the door and, if airborne material is involved, sealing edges with tape.

Once confinement has been accomplished, the cleanup can be conducted later. A well-constructed plan might be as follows.

8.8.7.2 Area Decontamination

A specific plan must be worked out considering the physical facilities and the properties of the contaminating materials. Such a plan should include:

1.   monitoring the contaminated area to determine the extent of contamination and the hazard;

2.   making sure decontamination personnel wear sufficient protective clothing; and

3.   in the case of liquid spills, proceeding with decontamination by scrubbing surfaces with a detergent solution, always working toward the center of the contaminated area, taking care not to spread the contamination to less active areas.

During the decontamination procedure, frequent and through monitoring, especially of all personnel and materials, should be performed before either is permitted to move into clean (uncontaminated) areas.

8.8.7.3 Personnel Decontamination

When a patient or a radiation worker becomes or is suspected of being contaminated, certain measures should be taken immediately. A G-M tube or other suitable detector of high sensitivity should be used to locate the area of contamination. Any contaminated articles of clothing, including shoes, should be immediately removed and placed in a lead-shielded storage area until decayed to a safe level. This period of time is usually 7 to 10 half-lives of the radionuclide spilled, but may vary depending on the amount of activity involved and the specific radionuclide(s).

When the hands or other body surfaces have become contaminated, care must be taken to prevent the spread of the radioactive material to other areas, especially to open wounds. Loose particles of contamination should be immediately removed. Decontamination methods that spread localized material or increase penetration of the skin should be avoided.

Decontamination of wounds should be accomplished under the supervision of a physician when possible. An effective method is to irrigate the wound(s) profusely with tepid water and to scrub gently with a soft brush, if necessary. It is important to avoid the use of hot water, which increases the blood flow to the skin surface; or of highly alkaline soap, which may fixate the contaminant; or of organic solvents, which may increase skin penetration by the contaminant.

For intact skin, the decontamination rules are similar. Wet the skin thoroughly and apply detergent. Work up a full lather, and keep it wet. Work the lather into the contaminated area by rubbing gently for an extended period, at least 3 minutes, while at the same time tepid–to–cold water is frequently applied. Rinse the area thoroughly with tepid–to–cold water, trying to limit run-off water to the contaminated areas. In the case of the fingers and toes, if monitoring shows excess radioactivity at the tips, it will help to clip the appropriate nails. If monitoring continues to show that the intact skin areas are highly radioactive, wash again, scrubbing with a soft brush, if necessary. If the radiation level still persists, help should be sought from someone knowledgeable in the area of radiation protection.

8.9 Review Questions

1. Who has the responsibility of achieving appropriate radiation protection standards?

2. What are the major duties of the RSO?

3. What is meant by a controlled area and by a supervised area? In the medical field, is any distinction made between them?

4. What does the ICRP recommend concerning the training of radiation workers? of physicians?

5. What is the responsibility of the physician toward the patient with respect to procedures involving radiation? research involving radiation?

6. What criterion should be used in implementing radiation protection standards when a facility is designed?

7. Describe some quality assurance measures that pertain to your present (or future) work situation.

8. What are the allowable leakage values for X-ray generator source housing as recommended by the ICRP?

9. Why should beam limiting devices be used? What is recommended with respect to dental equipment?

10. What does the ICRP recommend with respect to filtration of general purpose X-ray beams? What about dental units? What about equipment that generates only very soft (low-energy) X-ray photons?

11. Why is it important to check that exposure timers, especially on fluoroscopic, mammographic and dental units are functioning properly?

12. What parameters should a good generator control panel display?

13. What is the best protection for a patient undergoing a procedure involving radiation?

14. Describe some protective devices to be worn by the radiation worker; to be used to protect the patient.

15. What is the best auxiliary equipment to use with fluoroscopy?

16. What is a sealed source of radiation? In what way(s) are sealed sources used?

17. What precautions should be taken in the care of patients who have received diagnostic quantities of radionuclides? oncologic quantities?

18. Describe some of the precautions recommended when a particle accelerator is used for radiation oncology.

19. Describe some of the precautions recommended when a neutron generator is used for diagnostic or oncologic applications.

20. What are some common external sources of irradiation?

21. How can one best protect oneself from these external sources of irradiation?

22. Why are internally deposited radionuclides hazardous?

23. Upon what properties of the particular radionuclide internally deposited does the amount of internal irradiation depend? Why are beta and alpha particles less of a threat when external and much more dangerous when deposited internally?

24. Why do photon emitters pose much less of a threat when deposited internally than an equal amount of an alpha or beta emitter would pose?

25. What are the implications of stating a maximum permissible concentration for a radionuclide in water or in air?

26. What are the common methods of entry of radionuclides into the body?

27. Where should preparations involving volatile liquids be done? Where should those involving fine powders be done?

28. Describe some methods for contamination control.

29. Describe some specific shielding devices used for controlling irradiation by unsealed sources of radionuclides.

30. Describe some protective devices and decontamination equipment used for controlling contamination when working with liquid radionuclides.

31. What protective devices are available for protecting against airborne radioactivity?

32. What principles should be used in the management of accidents involving radionuclides?

33. What general principles are involved in confinement and decontamination after an accident involving a liquid radionuclide?

34. What are the general principles involved in personnel decontamination?

8.10 Problems

1. An average person drinks between 1.5 and 2.0 liters (1500 to 2000 cc) of liquid a day. If all of the liquid that a radiation worker were to imbibe contained 2.22 Bq·cc^{-1} of ^{131}I how much radioactivity would be imbibed in a day? If all the radioactivity imbibed were to be deposited in the body, would this exceed the maximum permissible body burden? (Note: the amount of radioactivity constantly decreases through physical decay and bodily elimination.)

2. An average adult person not doing particularly heavy labor, inhales about 10 m^3 (which is 10·10^6 cc of air) in eight hours. If the air contains a concentration of 3.33·10^{-4} Bq·cc^{-1} of ^{131}I, how many Bq's of this radionuclide would a radiation worker inhale in five-day working week? (Note: this is not equivalent to the body burden because the deposition of a radionuclide which enters the body through the lungs is quite complex. See report of ICRP Task Group on Lung Dynamics, Health Physics, 12:173-207, 1966.)

Chapter 9

RADIONUCLIDES AND THE LAW

The production, transportation, possession, and use of radionuclides are strictly regulated by law. This regulation is an effort by society to prevent radiation accidents and to ensure proper corrective procedures should such accidents occur. In the United States as well as in Europe and Asia, a strict licensing system controls every phase of activity associated with radionuclides. These licenses are issued at national levels in most countries and at the federal, state, and local level in the United States, depending upon a number of different factors. In general, these licenses include regulations that require a certain amount of expertise on the part of the recipient and some record keeping.

In the radiologic and health sciences, the technical and administrative measures involved in complying with the law are relatively minor. In the nuclear power industry, the regulations are lengthy and exacting. Nevertheless, whether the compliance demands are large or small, they should be attended to with care, diligence, and perseverance.

9.1 Licensing

In most countries the regulations governing the production, receipt, transportation, use and disposal of (sealed and unsealed) radionuclides are drawn from the recommendations of the ICRP, and from the publications of the International Atomic Energy Commission (IAEA) and the United Nations World Health Organization (WHO). For example, the five Nordic countries of Denmark, Finland, Iceland, Norway, and Sweden have agreed together on a set of recommendations to serve as a platform for more formal rule making in those countries.

Through the Atomic Energy Act of 1954, the United States Congress authorized the United States federal government to control reactor-produced (i.e., fission-produced) radionuclides, usually referred to as by-product material. This act provided for programs to develop atomic energy for the general welfare and also to regulate its use to protect public health and safety. The disadvantage of this system is that the formulating of regulations is under executive, rather than legislative, control. Because only fission-produced radionuclides were regulated at the federal level, the various state and local governments in the United States made rules regarding all other radioactive material. However, in 2005, the United State Congress authorized the regulation of accelerator produced radioactive material as well as radiation producing machines at the federal level. We will describe in some detail the regulations in the United States, as they are very similar to those in most other nations.

9.1.1 Nuclear Regulatory Commission

In the United States, the regulatory power created by the Atomic Energy Act is presently vested in the Nuclear Regulatory Commission (NRC), formerly the Atomic Energy Commission (AEC). The Code of Federal Regulations (CFR) is that set of laws enacted by federal agencies empowered by Congress to make regulations. Title 10 of the CFR pertains to atomic energy. The most important sections for radionuclide users are:

1. Notices, Instructions, and Reports to Workers: Inspection and Investigations (Part 19).

2. Standards for Protection Against Radiation (Part 20).

3. Rules of General Applicability to Domestic Licensing of By-product Material (Part 30).

4. Medical Use of By-product Material (Part 35).

Radionuclides such as a discrete source of any naturally occurring radioactive material. e.g., ^{226}Ra, that are not reactor-produced, or those that are produced in a particle accelerator such as a cyclotron are now (as of 2005) regulated by the NRC. The rule making process by the NRC includes all radionuclides produced intentionally or incidentally in an accelerator that is operated to supply radioactive materials for commercial, medical, or research activities. The NRC does not intend to regulate (as of 2006) the possession or use of particle accelerators, nor the incidental radiation produced when such machines are operated primarily for particle beam generation (including high energy physics and radiotherapy applications).

What is more, as mentioned in Chapter 4, in certain instances the NRC has entered into agreements with certain states such that the states themselves regulate all radioactive products. In these cases state regulations may be more, but must not be less, stringent than those of the NRC. In turn, states may enter into agreements with certain city or local governments to allow local regulation of radionuclides. Again, the same rule of more, but not less, stringent regulations applies. An example of a state entering into agreement with one of its cities is that of the State of New York and New York City.

9.1.2 Radiation Control for Health and Safety Act of 1968

It is well to note that radionuclides produced for medical use have been strictly regulated since they became available in a practical manner following World War II. The same is not true of X-rays or of other types of radiation because their use became very widespread before their possible harmful effects became known. The United States Radiation Control for Health and Safety Act of 1968 was the first piece of legislation in that country regulating both ionizing and nonionizing electromagnetic and particulate radiation as well as sonic, infrasonic, and ultrasonic emissions from electronic products. Many other countries have adopted similar regulations in this same time frame. In the 1968 Act, the United States Congress directed that the United States Department of Health and Human Services (formerly, Health, Education, and Welfare) develop and administer performance standards to control the emission of radiation from electronic products such as microwave ovens and color TV sets. It also authorized research by public and private organizations into the effects and control of such radiation emissions. The regulation of these products is usually accomplished through the Center for Devices and Radiological Health/FDA (formerly the Bureau of Radiological Health). Unfortunately, the implementation of this act has been hampered by many industrial concerns and "grandfather" clauses, but at least safety standards have been established and publicized. In 1990 the United States Congress passed an extension to 1968 act called the Safe Medical Devices Act which has been very helpful. Enforcement of the safety recommendations and regulations is difficult, but when a worker's safety is involved, sometimes the Department of Labor, which enforces the Occupational Safety and Health Act (OSHA) (Title 29 of the CFR), can be of help. Additionally, the 2005 act extends the authority of the NRC to include any radionuclides which the commission, along with other federal officials, determines may pose a health or safety threat or is used in commercial, medical, or research activities.

Additionally, the United States has formulated the "Mammography Quality Standards Act of 1992" (MQSA) which proposes standards for X-ray tube emission and image quality for diagnostic, particularly, screening mammography installations. The individual states are required to pass legislation which would further the aims of this act by requiring inspection and certification of mammographic units. The American College of Radiology (ACR) has established a procedure for a mammography installation to be certified by them as conforming with all the applicable regulations. This is enforced by the insurance companies who only reimburse a patient or institution if the facility has ACR accreditation. This is an ongoing process which has been extended to nuclear medicine and other diagnostic radiology procedures.

9.1.3 Authorization

When a program of work with radionuclides is initiated, authorization must be obtained from the appropriate government agency. The licensing mechanism in the United States will be described, although it is quite similar in all countries. In the United States, the NRC has decided that radionuclides can be obtained and used only under specific or general NRC licenses. Certain exceptions may be made in terms of particular items, concentrations, and quantities such as those used in academic physics laboratories for teaching purposes, or depleted uranium to be used in shielding. Since these exceptions rarely apply to the working situation of any radiation workers, they are not listed here.

A specific license is one issued to a named applicant after the commission has reviewed and approved the proposer's application. This application must include the proposed radionuclides and the pharmaceutical forms (if applicable), their intended use and maximum quantities to be possessed at any one time, the training and experience of the individual users, the radiation-detection instruments available, and the personnel monitoring procedures. In addition, a description of the laboratory facilities, the handling of equipment, and the waste-deposal procedures, as well as the radiation-protection program, is required.

If radioactive material is to be administered to humans, a specific license issued only to physicians is required. Title 10, Part 35.11-19 of the CFR states that an application for a specific license for diagnostic use of by-product material for uptake, dilution, excretion, or scanning procedures will be considered as an application for all of the diagnostic uses listed in 35.100 if the applicant meets the requirements of proper training and experience, demonstrates availability of appropriate equipment for performance of the procedures, and has appropriate monitoring equipment, waste-disposal procedures, radiation-protection program, and access to hospital facilities. The NRC in agreement with the United States Food and Drug Administration (FDA) regulates which radionuclides and pharmaceutical forms are listed in 35.100. If a physician wishes to use either a radionuclide or a pharmaceutical form not so listed, he/she must apply individually for each item.

An individual who plans to work with radionuclides at any institution such as a hospital must apply for an NRC (or other agency, such as city or state) license through that institution. Clinical programs sponsored by medical institutions are subject to a review by a radioactive materials committee, which must approve the license application. It is still true that a license is issued to an individual physician and must be specifically obtained for each radionuclide and for each pharmaceutical form not named in 35.100. However, physicians in training may participate in an institutional program. The institution, however, is responsible for the individual's compliance with provisions of the license.

Institutions that meet the requirements regarding staffing and facilities may obtain a license of broad scope. This license enables them to establish a radioactive materials committee, which may authorize specific members of their staff to work with radionuclides without special application to the regulatory agency for each individual user. Broad licenses can be issued for work involving animals only or for both animal and human use.

A distinction is made regarding the use of radioactive materials in a live person, *in vivo*, and the use of such materials in a laboratory test involving tissues or fluids extracted from a living person, *in vitro*. Licenses may be issued to use certain radionuclides, such as ^{125}I, *in vitro*, but not *in vivo*. Part 35 of Title 10 governs only *in vivo* uses.

Any institution that uses radionuclides must have a radiation-safety program conducted under the authority of a radioactive materials committee and implemented by a radiation safety officer (RSO, see Chapter 8 Section 1). The duties of the RSO reach into every aspect of radiation safety. To name but a few, the RSO prepares regulations, advises on matters of radiation protection, maintains a system of accountability for all radioactive material from procurement to disposal, inspects work spaces and handling procedures, determines personnel radiation exposures, monitors environmental radiation levels, and institutes corrective action in the event of accidents or emergencies. Since all users of radionuclides are subject to periodic inspections and committee reviews, the RSOs duties are of the utmost importance.

Certain general licenses are issued to any physician for specific diagnostic uses of specially prepared radiopharmaceuticals in prepackaged individual doses. Examples of these are ^{131}I as sodium iodide for measurement of thyroid uptake, or as iodinated human serum albumin (I-HSA) for determination of blood and blood-plasma volume. The physician must fill out a registration form stating that he/she has and is competent to operate appropriate radiation measuring instruments. The authorization is valid only when he/she has received a properly stamped copy of his registration form from the NRC. At no time does the general license authorize a physician to administer radiopharmaceuticals to a woman confirmed pregnant or to a person under 18 years of age. The maximum levels of activity that the physician may possess at any one time are limited to 7.4 MBq of ^{131}I, ^{125}I, and chromium-51 (^{51}Cr) and to 0.2 MBq of cobalt-58 (^{58}Co) and ^{60}Co. These general license holders are also exempt from the radiation-protection standards directed to other licensees. Last, general licenses are also issued by the NRC for use of ^{131}I or ^{125}I for *in vitro* clinical or laboratory testing.

9.2 Record Keeping

Once a person has received authorization to use radionuclides, that person becomes directly responsible for compliance with all the regulations governing the safe use of the radionuclides in his/her possession. The authorized person is also responsible for the safe use of those radionuclides by others who, under his/her supervision, work with the material.

A complete record must be kept on each radionuclide from its receipt through its final disposition. For this reason, in a large institution such as a hospital, the ordering and receiving of radionuclides are usually done through some centralized system overseen by the RSO. Each licensee receives the radionuclides from the central receiving area. Each authorized user must keep a complete record of how the radionuclide is used.

The amount of radioactivity administered to patients must be carefully noted together with the pharmaceutical form used. A separate master record is kept for each radionuclide. In this record, amount received, amount used for particular purposes and that purpose – e.g., for a particular patient study, for a particular animal experiment, for a specific laboratory application – and amount disposed of are carefully noted. At all times, an inventory of the amount of radioactive material on hand must be kept, and this inventory must be ready to be submitted to inspectors upon request.

When a generator system is used for obtaining a radionuclide, records must also be kept of the chemical and radioactive purity of the eluate (the generator product). For example, in the molybdenum–technetium (^{99}Mo–^{99m}Tc) generator system, it is necessary to check for and to record the purity of the product, ^{99m}Tc. Since the eluate should contain no ^{99}Mo, this procedure is usually known as checking for molybdenum breakthrough. The current regulations in the United States limit the amount of ^{99}Mo to 5.5 kBq (0.15 mCi) ^{99}Mo per 37 MBq (1 mCi) ^{99m}Tc.

In addition, this generator contains an alumina column. Since alumina is nonradioactive, its break-through must be tested for by chemical means. Alumina break-through is rare and is usually tested for by the manufacturer before the generator system is delivered and is not tested for on a routine basis by the end user.

Equally important, records must be kept of personnel exposure, radiation surveys or wipe tests, instrument calibration, waste disposal, and radiation incidents. In an institutional setting these records are usually maintained or at least overseen by the RSO. The records kept with respect to all radiation activities represent the main proof that an authorized user has of his/her compliance with the radiation-protection regulations. They are, consequently, important for legal purposes as well as for effective administration of the radiation-protection program.

9.3 Area Posting and Radionuclide Labeling

All countries which regulate the use of radiation in general and of radioactive material in particular, require that certain types of signs be used to warn of danger or of possible danger from the presence of radiation. The requirements include all areas where radiation in any form may be encountered and is not restricted only to areas where radionuclides are used. Warning signs are necessary because individuals might otherwise be unaware of the presence of radiation. At the same time, the sign used should reflect the actual radiation danger present and should not be more or less serious than needed. All signs must bear the three-bladed trefoil symbol of the exploding atom, which is the radioactive caution symbol. This symbol is usually colored magenta, black, or purple on a yellow background with the prescribed dimensions (Figure 9.1).

The types of signs and labels used are as follows:

1. **Restricted Area:** This type of sign may be used, but is not required. It is appropriate when a person might receive an effective dose in excess of 20 µSv per hour if he/she were continually present in the area, but not in excess of 50 µSv per hour. The area itself must be controlled by license.

2. **Caution: Radiation Area:** This sign is used, but again is not required when the personnel having access to the area may receive in any 1 hour an effective dose of 50 µSv or in any 5 consecutive days a effective dose in excess of 1.0 mSv to a major portion of the body. The area itself must be controlled by license. It is useful to post "Unauthorized Persons Prohibited" signs at the entrance to such areas.

3. **High Radiation Area:** This sign is required if a person in the area could receive an effective dose to a major portion of the body in excess of 1.0 mSv in any 1 hour. Such areas also require audible or visible alarm signals, such as warning tones or flashing, colored lights. (Figure 9.2)

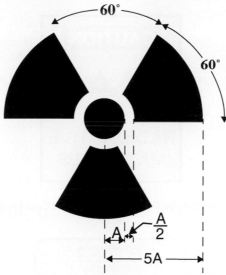

Figure 9.1: The three-bladed radioactive caution symbol with the required dimensions.

Figure 9.2: Caution sign required for a high radiation area. (Courtesy of Cone Instruments, Inc.)

4. **Caution: Radioactive material (room):** In general, rooms in which radioactive materials are used and/or stored must be posted with this sign in addition to any others that may also be required. However, small quantities are exempted but these exemptions are usually not applicable in medical or industrial work. (Figure 9.3)

5. **Airborne Radioactivity Area:** This sign is required if airborne radioactivity exceeds at any time the maximum permissible concentrations listed in Table 8.1 for 40 hours of occupational exposure. It is also required if the average over the number of hours in any week during which individuals are in the area exceeds 25% of the maximum permissible concentration.

6. **Caution: Radioactive Material (label):** This label is required on any container in which quantities of radionuclides are transported, stored, or used. Even refrigerators and containers for waste disposal should be so labeled. When the containers are used for storage, the labels should state, additionally, the nuclide(s) present, the amount of activity of each nuclide, and the date of each assay. (Figure 9.4)

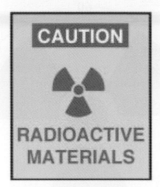

Figure 9.3: The radioactive materials sign for a room. (Courtesy of Cone Instruments, Inc.)

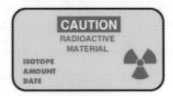

Figure 9.4: The radioactive material label for containers. There is space to fill in the radionuclide used, the amount of material and the date of the calibration. (Courtesy of Cone Instruments, Inc.)

9.4 Storage and Delivery

Ideally, radionuclides should be delivered to a central location where the contents of the package are inspected, monitored, logged, and stored by trained personnel until the material is picked up by the authorized user. Upon receipt, all containers of radionuclides should be carefully checked for evidence of leakage or breakage.

Should delivery be made to the general receiving area of an institution, the package should be logged in and quickly transferred to the user or to a storage area controlled by the RSO until it is delivered to or picked up by the licensee. Clearly visible signs giving specific instructions pertaining to the handling of radioactive material packages should be posted conspicuously in the receiving area. Records should also be kept of the names of the person receiving the package and the person to whom it is transferred or who places it in a locked storage area.

When radionuclides requiring a Radioactive Materials label are to be stored, the areas chosen must be locked and protected against fire, explosion, or flooding. Radionuclides must also be stored in adequately shielded containers. The practical rule is to shield stored materials so the radiation level is less than 50 μSv per hour at a distance of

1 foot from the shield's surface. It is highly desirable to keep the radiation level below 1.0 mSv per hour; otherwise, the storage area must be treated as a high-radiation area. The RSO must be informed of any transfer of radionuclides to a new area, i.e., an area previously undesignated for storage.

Whenever radionuclides are ordered, a copy of the user's license must be kept on file by the vendor. It is the vendor's responsibility to see that only those radionuclides covered by the user's license are delivered.

9.5 Transport

In the United States as in other countries, government regulations pertaining to packaging and transporting of radionuclides are myriad and confusing. The following Unites States publications contain most of the regulations:

1. Title 49 Parts 170 to 199 of the CFR, administered by the Department of Transportation (DOT), applies to packaging and shipment of radioactive materials by rail, highway, or water transportation. For water transportation shipment regulations, see also Title 46 Parts 146 and 149 of the CFR.

2. Title 14 Part 103 of the CFR regulates shipments via air transportation.

3. Title 10 Part 71 of the CFR, administered by the NRC, regulates certain shipments of fissile and large quantities of NRC licensed material. The NRC also regulates the export and import of by-product material through Title 10 Part 36 of the CFR.

4. Title 39 Parts 124.2(d) and 125(d) regulate shipment of radioactive materials using the United States Postal Service.

In addition, agreement states and, in the case of New York City, agreement cities have added further regulations to those already in existence. In Europe and other parts of the world, local regulations pertaining to the transportation of radionuclides are based on the recommendations of the IAEA stated in the publication *Regulations for the Safe Transport of Radioactive Materials*. These recommendations are very similar to the regulations in force in the United States.

It is possible in most countries (Norway is an exception), but under stringent circumstances, to send radionuclides through the postal service. The brief summary of mailing regulations as contained in the United States Postal Service pamphlet, Radioactive Matter, Publication 6, April, 1971 is a typical example of those in force everywhere. It is more usual to send radionuclides via a carrier licensed specifically for transporting them.

In the United States, if the cargo-carrying vehicle is not used exclusively for transporting radionuclides, the effective dose rate must be less than 2.0 mSv per hour at the surface of the package and 0.1 mSv per hour at 3 feet from the package. If, however, the transporting is done by a vehicle used exclusively for this purpose and is loaded and unloaded by properly trained personnel, the effective dose rate can be as high as 10 mSv per hour at 3 feet from surface of the package, 2.0 mSv per hour at any point on the external surface of the vehicle, 0.10 mSv per hour at 6 feet from the external surface of the vehicle, and 0.02 mSv per hour in any normally occupied position in the vehicle.

Licensed users who send radionuclides by taxi or other vehicles that primarily carry passengers must check that the carrier has the appropriate commercial license for transporting radioactive materials. Proper shipping papers and other record-keeping requirements should also be observed. If a user is licensed to use his own car for transporting radionuclides, his insurance policy should be checked for possible exclusion clauses with regard to accidents involving radioactive materials.

If radionuclides are carried on foot from one institution to another, proper precautions, such as the use of shatterproof containers and adequate shielding, must be observed. The effective dose rate at 3 feet from the container must be less than 0.1 mSv per hour, and tests must be made to determine that no removable contamination exists on the surface of the container. When such material is transported by foot messenger, it should be routed to encounter the minimum number of pedestrians.

Packages containing radionuclides in amounts that are not mailable must be labeled according to the following method. The labels, which are diamond-shaped and bear radioactive caution signs, are internationally uniform. In all, three different labels, as shown in Figure 9.2, are used as follows:

1.   **radioactive white I**; this label is all white with one red stripe in the lower half of the diamond. It is used on packages where the effective dose rate at any point on the external surface is less than 5 μSv per hour;

2.   **radioactive yellow II**; here the upper half of the label is yellow; the lower is white and contains two red stripes. This label is used on packages where the effective dose rate is between 5 and 100 μSv per hour on the surface and less than 5 μSv per hour at 3 feet according to the regulations in force in the United States; the IAEA recommends that the effective dose rate not exceed 50 mSv per hour at any location on the external surface of the package and that the transport index (see Glossary) be one.

3.   **radioactive yellow III**; this label is yellow on the upper half of the diamond and white below. It contains three red stripes in the white half. In the Unites States this label signifies that the effective dose rate at the surface of the package is greater then 100 μSv per hour or is greater than 5 μSv per hour at 3 feet. The effective dose rate, however, cannot exceed 500 μSv per hour at 3 feet. The IAEA recommends that the effective dose rate not exceed 2 mSv per hour at any location on the external surface of the package and that the transport index be 10. Vehicles transporting packages with radioactive yellow III labels must display a radioactive sign consisting of the three-bladed radioactive caution symbol coupled with the words "Radioactive Materials".

Furthermore, each radionuclide is given a transport group designation, and each aggregate shipment is given a fissile class and a transport index (see Glossary). These

Figure 9.5: Radioactive caution labels for the transport of radionuclides. (a) Radioactive White I: the label is all white with a black three-bladed radioactive caution symbol in the upper half and one red stripe in the lower half. (b) Radioactive Yellow II: the upper half is yellow with a black three-bladed radioactive caution symbol in the upper half; the lower half is white with two red stripes. (c) Radioactive yellow III: the upper half is yellow with a black three-bladed radioactive caution symbol in the upper half; the lower half is white with three red stripes.(Courtesy of Cone Instruments, Inc.)

concepts are defined in Title 49 Part 170 and 173 of the United States CFR and in the above mentioned IAEA document.

9.6 Radionuclides and Animals

When animals are injected with radionuclides, contamination-control procedures should always be employed. One common method is to use trays lined with absorbent material. The cages of such animals must have labels stating the radionuclide, quantity of material injected per animal, date of injection, and name of the licensed user. Special cages, such as metabolic-type cages, should be used any time contamination is a problem. The radioactive animals should be housed separately from other animals.

It is important that adequate ventilation be provided when a radioactive substance in animals kept after injection may become volatile and hence dispersed into the area housing them. Animal handlers should be instructed by the authorized investigator with respect to exposure levels, time limitations in the area, and handling requirements of animal carcasses and excreta.

If the animal excreta are not mixed with sawdust or wood-shavings and are below the limits given in Table 8.1, they may be disposed of in the sewerage system. Otherwise, they should be placed in plastic bags and disposed of as solid waste. The animal carcasses should be dealt with as solid waste.

9.7 Waste Disposal

Radioactive waste is inescapable when unsealed radionuclides are used. Whether the waste results from medical uses, research or industrial applications, it is necessary to manage it in such a way that while the radioactivity is decaying to negligible levels there will be no excessive exposure of radiation workers or of the general public. Waste disposal is always under the direction of the RSO. In any case to avoid accumulation of waste in working areas, it is necessary to have a system of collection and transfer to a storage area prior to disposal. Containers for waste disposal were described in Chapter 8. These containers must always be clearly labeled as containing radioactive materials. Typical detailed instructions for disposal of radioactive materials are contained in Title 10 Parts 20.2001–2007 of the Unites States CFR. Radionuclides, particularly common in medical and research activities, having a rather short half-life, e.g., less than a month, can be stored for a relatively short period until the activity has been reduced to that of background activity. The United States NRC permits decay-in-storage before disposal in ordinary trash if the by-product material has a physical half-life of less than 65 days and if:

1. the by-product material is held for decay a minimum of ten half-lives;

2. by-product material is monitored at the container surface before disposal as ordinary trash and it is determined that its radioactivity cannot be distinguished from the background radiation level with a radiation detection survey meter set on its most sensitive scale and with no interposed shielding;

3. all radiation labels are removed or obliterated; and

4. each generator column is individually separated and monitored with all radiation shielding removed to ensure that it has decayed to background radiation level before disposal.

For longer-lived radionuclides, the common methods of disposal can be classified as dispersion, either through air or through water or both, and concentration into solid waste.

Gaseous radioactive waste can arise from chemical or fission reaction, from diagnostic procedures which use gases such as ^{81}Kr and ^{133}Xe, or from use of volatile liquids such as iodine. Chemical fume hoods used with gaseous radioactive sources should be vented directly into the atmosphere. Care must be taken, however, that the amount of radioactivity vented into the atmosphere does not exceed, when diluted with air, the MPC as defined in Chapter 8 for radiation workers or for the general public. It must also be assured that it does not re-enter the premises through a window or ventilating system. For activities too high for direct disposal, traps and/or filters must be used to contain the radioactivity for later disposal as solid waste.

Liquid radioactive wastes that arise from diagnostic or oncologic procedures, residues of stock radionuclides, chemical or biological experiments, reactor coolants, decontamination procedures, patient excreta or vomit, can be dispersed into the sewage or other appropriate water system providing that when diluted with water the MPC as defined in Chapter 8 is not exceeded. In the case of solutions or organic substances such as toluene, ethanol and dioxan, frequently used in medical applications, the chemical toxicity as well as the radioactivity level must be considered. If it is impractical because of its high activity and/or chemical toxicity to store the liquid waste for later disposal as a liquid, the ordinary method is to solidify the liquid waste with resin or concrete and to dispose of it as solid waste.

Solid radioactive wastes arise from many sources including chemical precipitates and animal carcasses. While incineration is permitted in some countries, it is discouraged. Incineration results not only in dispersal of vapor and fine ash into the atmosphere, but residual radioactivity in the furnace ash and flue. Each of these presents a hazard. In general, solid wastes are stored in properly shielded and designated areas until sufficiently decayed to allow disposal as ordinary waste. In the case of long-lived radionuclides, when on-site decay is not possible, the complex generating the waste either has a burial site and procedures strictly regulated by law, or hires an outside concern licensed to dispose of waste. If an outside vendor is employed, that company is responsible for collection from the central site and disposal of the radioactive waste material. The vendor is strictly regulated by law and is responsible for safe burial of the refuse. All waste given for burial must be clearly marked as to type and quantity of its chemical as well as its radionuclide contents. Human cadavers containing radionuclides should be inspected by the RSO before the body is removed to the morgue. If it is determined that a radiation hazard exists, a "radioactive materials" tag should be attached to the body and appropriate instructions issued to the funeral director or the hospital morgue.

9.8 Review Questions

1. With respect to radionuclides, what is strictly regulated by law?

2. How is this accomplished in the United States and by whom?

3. What agency is charged with the enforcement of the United States federal regulations? Where are these regulations published?

4. In the United States, how are radionuclides that are not under federal control regulated? Under what circumstances can the federal agency named above review their use?

5. Is there a need for controlling radiation sources other than radionuclides? How is this done?

6. Describe the particular types of licenses for possession and human use of radionuclides issued by the United States NRC or by another proper authority.

7. Describe the type of records that must be kept in conjunction with approval to possess and use radionuclides.

8. What type of area posting and radionuclide labeling is recommended or required?

9. Describe the regulations governing the storage and delivery of radionuclides.

10. What agencies govern the transport of radionuclides?

11. List some of the regulations governing the transport of radionuclides.

12. Describe the three types of labels to be used on packages containing radionuclides in amounts that are not mailable.

13. With respect to disposal of radioactive waste, when may radionuclides be stored for ten half-lives and then disposed of as ordinary waste?

14. Describe the precautions necessary when disposing of gaseous and liquid radioactive waste.

15. Describe the procedures required for the disposal of solid radioactive wastes.

16. What should be done if a patient containing radionuclidic material dies?

9.9 Problems

1. One liter (1000 cc) of liquid containing 4 kBq of ^{125}I is to be disposed of as liquid waste. How many liters of water must this be diluted into (total volume) so as not to exceed the MPC of ^{125}I in water. Consider the water to be in an unrestricted area.

2. One liter volume of ^{133}Xe is determined to have an activity of 10 kBq and is to be dispersed into the atmosphere. The vent flue available for this procedure has an opening of one meter by one meter. If the air is moving pass the vent at the rate of 10 m·s^{-1}, at what linear rate should the Xe be released into the air so as not to exceed the MPC? Consider the air to be in an unrestricted area.

3. A package containing radionuclides is to be shipped by special transportation. If the effective dose rate at the surface of the package is 50 µSv per hour, but less than 5 µSv at three feet from the surface of the package, what type of radioactive caution label should be used?

Chapter 10

SHIELDING FROM EXTERNAL RADIATION

The establishment of effective/equivalent dose limits for radiation workers as well as for the general public allows calculation of shielding requirements for walls, floors, and ceilings of rooms that contain unsealed sources of radiation, diagnostic or radiation therapy X-ray machines, sealed-source therapy units, or linear accelerators. The thickness of radionuclide containers can also be determined. Proper shielding design, or barriers, for areas in which radiation is present allows use of adjacent areas for other activities. These adjacent areas are designated as either controlled or uncontrolled.

A controlled area is one habitually occupied by persons who are radiation workers and is under the jurisdiction of the RSO (see Chapter 8, Section 1). An uncontrolled area is one normally occupied by persons who are not radiation workers: e.g., the secretary's office and patients' waiting room. However, a radiologist's office and a technologists' lounge, provided they are within those areas controlled by the RSO, are considered controlled areas and are also considered occupied by personnel who are allowed the effective/equivalent dose limits for radiation workers. The ALARA principle (economic and social factors taken into consideration), which is normally 1/10 of the allowable limit for the area, is the present acceptable world-wide standard for both controlled and uncontrolled areas.

The limits for air kerma (exposure) allowed in any area are fixed by the effective/equivalent dose limits for the type of individual concerned. A radiation worker is allowed a effective dose limit of 50 mSv per year. If this effective dose limit is divided by 50, the approximate number of weeks in a year, the effective dose limit becomes 1 mSv per week. Similarly, the effective dose limit for a non-radiation worker who is infrequently exposed (5 mSv per year) is found to be 0.1 mSv per week and for a frequently exposed non-radiation worker (1 mSv per year) the limit is 0.02 mSv per week. Expressed in the former system of units, the air kerma limits are 0.1 rem per week for controlled areas and 0.01 rem (or 0.002 rem) per week for uncontrolled areas.

In the radiologic and health fields, most of the absorbed dose obtained is due to irradiation by photons. Since the quality factor for photons is equal to 1.0, the effective/equivalent dose limits may be written in terms of air kerma (see Chapter 3) as 1 mGy per week for controlled areas and 0.1 mGy (or 0.02 mGy) per week for uncontrolled areas. Expressed in the former system of exposure units, the limits are 0.1 R per week for controlled areas and 0.01 R (or 0.002 R) per week for uncontrolled areas. Following the ALARA principle, NCRP 147 (2005) recommends an target goal of 0.1 mGy per week for controlled areas. This would permit a pregnant worker to go about her normal duties.

Protective barriers in the form of added shielding consist of many different types. The thickness and composition of walls, floors, and ceilings of rooms designed to protect people in adjacent rooms-including those above and below-from radiation air kerma due to unsealed sources of γ–ray photons require one type of calculation, where as X-ray photons produced in diagnostic or radiation oncology X-ray machines or particle accelerators and annihilation photons produced by positron emitters require another, and sealed sources used in therapy machines or in brachytherapy require yet another. Additionally, it is important to understand the role of collimation in designing protective barriers. Both ICRP and NCRP have written recommendations in the form of reports (listed in the bibliography) that deal with protection from X-ray and gamma-ray sources as well as with protection in the dental and veterinary medical fields.

10.1 Protection from External Unsealed γ-ray Photons

This discussion has been confined to shielding from photons, as a few millimeters of aluminum is usually sufficient to shield from all but the highest-energy beta particles. Remember that when shielding from beta particles, the shielding material must be carefully chosen so that no photons are produced by bremsstrahlung, as discussed in Chapter 8, Section 8.8.1.

A specific radionuclide may emit more than one photon. Each photon is emitted with a fixed energy and at a fixed rate. For example, ^{131}I decays via beta decay to one of several excited nuclear states of its offspring, Xenon-131 (^{131}Xe). These excited nuclear states lose energy by emitting photons in a variety of ways, which produce photons of many different energies. A photon with an energy of 364 keV is produced 84% of the time. However, taking into account their respective decay probabilities, all other photons produced must be shielded against as well.

In order to account for all photons emitted by a radionuclide, together with the probability of decay of each photon, a constant known as the specific gamma-ray constant Γ is used. This constant describes air kerma rate due to photon emissions of a nuclide at a unit distance from the source; the source is considered to be a point. In the new system of units Γ is stated in $\mu Gy \cdot m^2 \cdot (GBq \cdot h)^{-1}$ whereas in the old system of units Γ was expressed as $R \cdot cm^2 \cdot (mCi \cdot h)^{-1}$.

Table 10.1 gives values of Γ for some commonly used radionuclides. The point-source assumption is valid provided the source volume is small compared to the distance from the source to that point at which air kerma rate is being measured.

The air kerma rate then for a specified activity of a particular radionuclide at a fixed reference point is given by:

$$X_r = \frac{\Gamma A}{d^2} \qquad (10.1)$$

where X_r is the air kerma rate, Γ is the specific gamma-ray constant for that particular radionuclide, A is the activity of the radionuclide at the time of measurement, and d is the distance in meters from the source to the position where air kerma is being measured.

Table 10.1: Specific Gamma-Ray Constant Γ for Some Commonly Used Radionuclides

Radionuclide	Γ $\dfrac{\mu Gy \cdot m^2}{h \cdot GBq}$	Γ $\dfrac{R \cdot cm^2}{h \cdot mCi}$
^{18}F	124.9	5.29[a]
^{51}Cr	3.8	0.16[b]
^{60}Co	311.6	13.20[b]
^{67}Ga	26.0	1.10[b]
^{123}I	35.4	1.50[a]
^{125}I	16.5	0.70[b]
^{131}I	51.9	2.20[a]
^{59}Fe	151.1	6.40[b]
^{75}Se	47.2	2.00[b]
^{82}Rb	137.6	5.83[a]
^{99m}Tc	16.5	0.70[c]
^{111}In	33.0	1.40[c]
^{133}Xe	2.4	0.10[c]
^{226}Ra	194.7	8.25[c]

[a]AAPM Report 108, AAPM, College Park, MD. (2005)
[b]Jaeger, R.C., et al.: *Engineering Compendium on Radiation Shielding*, Vol. 1. Berlin, Germany: Springer-Verlag, 1968, pp. 21–30. Values from Jaeger are reproduced with permission of the U.S. Department of Health, Education and Welfare, Public Health Service, Food and Drug Administration, Bureau of Radiological Health, from the Radiological Health Handbook, rev. ed., January, 1970.
[c]Oakridge Dosimetry Laboratory (personal communication).

In Chapter 7 the concept of HVL was thoroughly discussed. This concept is now used to determine air kerma rate from a radionuclide. From Eq. (7.13), replacing number of photons by air kerma rate:

$$\frac{X_r}{X_{rb}} = 2^{t/t_{1/2}} = 2^n \tag{10.2}$$

Here the air kerma rate with shielding present is written as X_{rb}. If X_{rb} is then made equal to X_{rp}, the maximum permissible air kerma rate for the area under consideration, i.e., controlled or uncontrolled, then the number of HVL's of shielding material necessary can be found by using Table 7.1.

Eq. (10.2) can be rewritten as follows, substituting for the value of X_r from Eq. (10.1):

$$\frac{\Gamma A}{X_{rb} \, d^2} = 2^n \tag{10.3}$$

and rearranging:

$$X_{rb} = \frac{\Gamma A}{d^2} \cdot \frac{1}{2^n} \tag{10.4}$$

The quantity 2^n is sometimes known as the attenuation factor, meaning that it represents the fraction of radiation transmitted by a protective barrier between the source and the place where air kerma rate is being measured.

The activity of a radionuclide decreases with time, so it is necessary to know the activity at the particular time for which the air kerma rate is being calculated. If air kerma is to be calculated for a given time period, e.g., for a working week of 40 hours, Eq. (10.4) can be simply multiplied by the time only if the physical half-life of the radionuclide is significantly longer than 1 week. If this is not the case, the nuclide must be decayed appropriately for each day or hour, and total air kerma obtained as a sum over suitable time periods. Air kerma can be written as:

$$X_b = \frac{\Gamma At}{d^2} \cdot \frac{1}{2^n} \tag{10.5}$$

where X_b is the air kerma for a time t, which is short compared to the radiological (physical) half-life $t_{1/2p}$ of the radionuclide. The subscript "b" is used to designate that a barrier is in place. If no barrier exists, $n = 0$ and $\frac{1}{2^n} = 1.0$. Then $X_b = X$, the air kerma when no barrier is in place.

There are actually two considerations with respect to shielding of unsealed γ-ray photons. The first is the shielding of all the areas where radioactive patients are present. This includes the injection area, the waiting area, and the actual scanning rooms. The second is the shielding for radionuclides which are being stored for use or disposal. The principles elucidated above apply mainly to the storage of radioactive material.

In the first case, the facility must be designed so the rules governing occupational and non-occupational areas are observed. For the usual radionuclides used in diagnostic procedures (excluding positron emitters), calculations should be done using the same methods as described in Section 10.2. In general, there is usually very little, 1/16 inch (1.6 mm) lead or equivalent, or no extra shielding needed in the scanning rooms to fulfill the ALARA principle.   However, the walls between adjacent scanning rooms generally need some extra shielding to prevent "cross -talk" between the γ-ray detection equipment, i.e., the radioactivity from a patient in one scanning room must not be observed by the detection equipment in an adjacent room.

The amount of shielding, whether for a small vial, a syringe, a storage compartment, or a patient area, will of course, depend on the radionuclide being used. This will be characterized by the gamma-ray constant for photons. If in addition to γ-ray photons, beta particles are present, then the shielding must be designed to eliminate any photons that might be produced from electrons by bremsstrahlung. The special case when the radionuclide is a positron emitter (annihilation photons are produced) will be considered below.

Example 10.1

A radiologist occupies a desk that is 0.5 m from a 100 MBq source of radium-226 (^{226}Ra), which has a physical half life of approximately 1600 years. How much shielding must be in place to reduce the radiologist's air kerma to the allowable 1 mGy per week? Γ for ^{226}Ra is 194.7 μGy·m^2·GBq^{-1} per hour, or $1.947\cdot10^{-4}$ mGy·m^2· MBq^{-1} per hour. Using Eq. (10.5), rearranging and substituting gives:

$$2^n = \frac{\Gamma At}{d^2 X_b} = \frac{1.947 \cdot 10^{-4}\frac{mGy \cdot m^2}{h \cdot MBq} \cdot 100\ MBq \cdot 40\frac{h}{week}}{0.25\ m^2 \cdot 1\frac{mGy}{week}}$$

and finally:

$$2^n = \frac{1.947 \cdot 4 \cdot 10^{-1}}{25 \cdot 10^{-2} \cdot 1} = 3.12$$

Therefore, from Table 7.1 two HVL's of material is required.

10.2 Protection from Sources of X-Ray and Annihilation Photons

X-ray machines and linear particle accelerators generally are not run continuously. Additionally, the patient, who is the source of radioactive air kerma is not continuously present in the detector. Consequently, the output of X-ray or annihilation photons is confined to certain definite time periods. Further, the intensity of the X-ray photons is a function of both tube current in mA and tube voltage in kVp or kVcp (see Glossary), or the accelerator energy. This is an important contributing factor because, as a rule, an X-ray beam travels in one primary direction, but may be scattered in many directions. In addition, the housing containing the X-ray generating apparatus is a source of leakage radiation. The intensity of the annihilation photons present is a function of the amount of radioactivity administered to the patient. For a complete discussion of shielding from annihilation photon see American Association of Physicists in Medicine (AAPM) Report 108 [AAPM 2005]. Below the principles will be discussed in relation to X-ray generators with a short discussion of annihilation photon protection at the end of the section.

10.2.1 Primary Barriers

In order to describe air kerma from the primary X-ray or annihilation photon beam, three standard parameters are used: weekly work load W, use factor U, and occupancy factor T. In addition, the energy spectrum (quality) of the photon beam must be considered.

10.2.1.1 Weekly Work Load W

Weekly work load W for a source of X-ray photons is the product of average current through the X-ray tube multiplied by maximum expected operating time for the generator in minutes per week. The units for W are, consequently, in mA-minutes per week. For annihilation photons, the number of patients imaged, the amount of radiotracer administered per patient, the length of time that each patient remains in the facility, and the method used to acquire data, i.e., using two-dimensional or three-dimensional geometry must be considered to determine the equivalent of workload.

10.2.1.2 Use Factor U

The ICRP has recommended values for U and T as follows: The value U = 1.0 designates full use and is used for the wall, floor, or ceiling routinely exposed to the primary X-ray beam. In most installations the primary X-ray beam points toward the floor; hence the floor of the X-ray room has a use factor of 1.0. The exception is dental installations, where the wall, floor or ceiling intersecting the primary beam has a use factor of 1/4. A use factor of U = 1/4 designates those walls of the X-ray room not routinely exposed to the primary beam. If the X-ray unit is used routinely in a fixed direction, say, vertically toward the floor, but may be used sometimes in a horizontal fashion (as a fixed or rotating unit), the walls of the room have a use factor of 1/4 (1/16 for dental installations). Finally, where rotation of the X-ray machine is possible, a value U = 1/16 indicates occasional use.

10.2.1.3 Occupancy Factor T

The occupancy factor T is based on the amount of time the specific area for which air kerma is being calculated is used. The ICRP recommends an occupancy factor of 1.0 for any space that is fully occupied. Any type of space in which people can congregate—a corridor big enough for a desk, an office, a children's play area—is a candidate for full occupancy. If the area is occupied for less than forty hours a week, the occupancy factor must be calculated assuming that occupancy of an area for forty hours a week is equivalent to an occupancy factor of 1.0.

On the other hand, an occupancy factor of 1/4 is applied to all areas that are partially occupied. Areas considered partially occupied are corridors within the department, lounge areas, utility rooms, patient's wards and rooms, elevators not routinely used by occupationally exposed personnel, and unattended parking lots. Finally, an occupancy of 1/16 is given to areas that are used only occasionally, for example, closets, stairways, automatic elevators, pavements, streets, and rest rooms not routinely used by occupationally exposed personnel.

10.2.1.4 Photon Energy

An additional consideration, with respect to reducing air kerma from primary X-ray photons, involves the energy of the photons themselves. As the kVp on an X-ray machine is raised, the quality of the X-ray beam changes, that is, the photons have a different energy spectrum. The HVL of the barrier used for shielding depends not only on the material composing the barrier but also on the energy of the photons impinging upon it. In order to account for different energies in the beam, another quantity K is defined. This quantity K gives air kerma in units of $mGy \cdot m^2 \cdot mA \cdot min^{-1}$ that is expected from a primary beam produced by a particular kVp. K takes into account X-ray tube housing and any other shielding that might be in place surrounding the X-ray tube itself.

Air kerma from photons produced by an X-ray machine may then be written as:

$$X_b = \frac{WUT}{d^2} K \qquad (10.6)$$

Here the symbol X_b is used to designate a barrier in place. The air kerma from X-ray machines must always be reduced to the maximum permissible level in order that areas near the radiation installation be safe for occupancy. Naturally, air kerma is always calculated in adjacent areas. Accordingly, X_b is always given a maximum value of 1 mGy, 0.1 mGy, or 0.02 mGy depending upon whether the area in which the air kerma is being measured is controlled or uncontrolled. Eq. (10.6) is usually written in terms of K as:

$$K = \frac{Pd^2}{WUT} \tag{10.7}$$

where P has been substituted for X_b and stands for the maximum permissible air kerma per week. The value of K is then found on graphs prepared by the ICRP or the NCRP. Figure 10.1 through Figure 10.3 are graphs of values of K versus thicknesses of lead, and Figure 10.4 is a graph of the values of K versus thicknesses of concrete. (These graphs are illustrative only. The primary references should be consulted for professional calculations.) After evaluating K, the proper amount of the designated material to be added to the appropriate wall, floor, or ceiling in order to reduce air kerma for a particular kVp value to the allowable limit is determined. Additionally, Table 10.2 lists the equivalent thickness of concrete that can be used instead of lead. These graphs and the table represent empirical data gathered by many different experimenters. The ICRP indicated that the data shown for constant potential generators may be used for all types of generators without introducing serious discrepancies, and because the beam filtration is negligibly small, they are dependent only on peak operating potential.

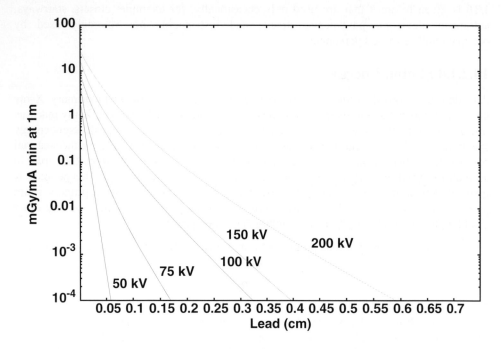

Figure 10.1: Broad–beam transmission of X-rays through lead, density 11,350 kg· m^{-3}. Constant potential generator; tungsten reflection target; 2 mm aluminum total beam filtration. Ordinate intercepts are 28.7 at 200 kV, 18.3 at 150, 9.6 at 100, 6.1 at 75, and 2.6 at 50. Note that the y-axis is a log scale. (Data for curves taken from ICRP Publication 33. Amsterdam, The Netherlands, Elsevier B.V. 1982.)

Example 10.2

A tomographic diagnostic X-ray machine is used at an average of 10 mA for 100 minutes each week at 150 kVp. The X-ray tube is positioned 4 meters from a wall between the radiation room and a radiologist's office which is occupied by the same person an average of 20 hours per week. The wall contains 4 centimeters of ordinary concrete. The primary X-ray beam is usually directed toward the wall adjoining the radiologist's office. What thickness of lead must be added to the wall? Using Eq. (10.7), where P equals 1 mGy since the radiologist's office is a controlled area; d is 4 m; W is 10 mA·100 minutes per week, which equals 1000 mA-minutes per week; U is 1/4; and lastly T is 20/40, or 1/2, the calculation can be written as:

$$K = \frac{Pd^2}{WUT} = \frac{1\frac{mGy}{week} \cdot 16\ m^2}{\frac{1000\ mA \cdot min}{week} \cdot \frac{1}{4} \cdot \frac{1}{2}} = 0.128 \frac{mGy \cdot m^2}{mA \cdot min}$$

From the 150 kVp curve in Figure 10.1, 1.2 millimeters of lead are required. The wall already contains 4.0 centimeters of concrete. From Table 10.2 it is found that 1.0 millimeter of lead corresponds to 8.0 centimeters of concrete. Since the wall contains 4.0 mm of concrete, the means that it has the lead equivalent of 0.5 mm. Since 1.2 mm are required, this means that 0.7 millimeters of lead, or its equivalent in concrete, must be added to the wall.

10.2.2 Scattered Radiation

It is just as important to shield against scattered radiation as against primary radiation. The ICRP refers to all photons emitted by an irradiated object as scattered radiation even though some of them are not due to Compton interactions. It is possible for radiation to be scattered to a location that is not exposed to the primary X-ray beam.

The primary source of scattered radiation always originates at the patient's position. All absorption by the patient of the primary X-ray photon beam is always neglected. The amount of radiation air kerma to a particular location due to scattering depends upon the quality of the X-ray beam (kVp value), the cross-sectional area of the X-ray beam at the scatterer, and the angle of scattering of the rays. The use factor U is always one for scattered radiation. For this case Eq. (10.7) is modified in the following ways:

$$K_s = \frac{100 \cdot P \cdot d^2}{W \cdot T \cdot S} \tag{10.8}$$

P and T are defined as above. The workload W is the same as before unless the distance between source and scatterer (usually the patient) is not one meter. If this distance is different from one meter, W must be modified by the inverse square law. For example, if the source–scatterer (skin) distance is 50 cm, the K obtained above must be multiplied by 1/4; if the source–scatterer distance is 70 cm, then K is multiplied by 1/2. The distance d is measured in meters from the source of scattered radiation to the point of interest. The value

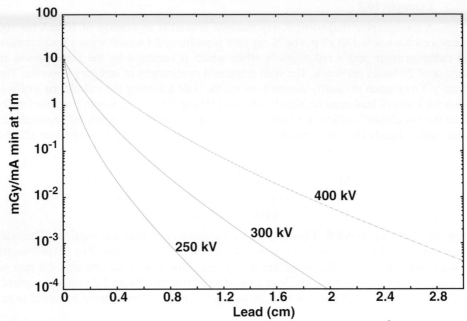

Figure 10.2: Broad–beam transmission of X-rays through lead, density 11.350 kg·m^{-3}; 250 kV: constant potential generator; tungsten reflection target; 0.5 mm copper total beam filtration. 300 and 400 kV: constant potential generator; gold reflection target; 3 mm copper total beam filtration. Ordinate intercepts are 23.5 at 400 kV, 11.3 at 300, and 16.5 at 250. Note that the y-axis is a log scale. (Data for curves taken from ICRP Publication 33. Amsterdam, The Netherlands, Elsevier B.V., 1982.)

Table 10.2: Equivalents Between Lead and Concrete[a] at Various X-ray Tube Potentials

Concrete equivalents (mm) for the Listed X-ray tube potentials				
Lead thickness (mm)	150 kVp	200 kVp	300 kVp	400 kVp
1	80	75	56	47
2	150	140	89	70
3	220	200	117	94
4	280	260	140	112
6	–	–	200	140
8	–	–	240	173
10	–	–	280	210
15	–	–	–	280

[a] Concrete taken as having a density of $2.35 \cdot 10^3 \text{ kg·m}^{-3}$. Reprinted with permission of the U.S. Department of Health, Education and Welfare, Public Health Service, Food and Drug Administration, Bureau of Radiological health, from the radiological Health Handbook, rev.ed., January, 1970

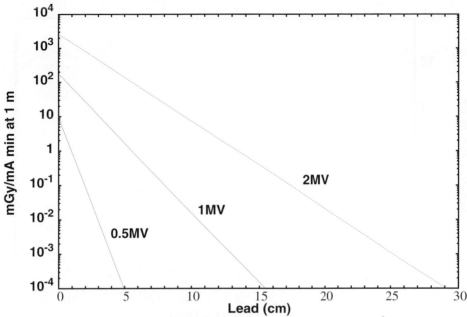

Figure 10.3: Broad–beam transmission of X-rays through lead, density 11,350 kg·m^{-3}. Constant potential generators. 0.5 and 1.0 MV: 2.8 mm tungsten transmission target followed by 2.8 mm copper, 18.7 mm water, and 2.1 mm brass beam filtration. 2 MV: high atomic number transmission target; 6.8 mm lead equivalent total beam filtration. Ordinate intercepts are 2610 at 2 MV, 174 at 1, and 1.9 at 0.5. Note that the y-axis is a log scale. (Data for curves taken from ICRP Publication 33. Amsterdam, TN, Elsevier B.V ,1982.)

of S is the percent (hence the factor 100 in the numerator) of the incident absorbed dose rate scattered to one meter. Table 10.3 gives some representative values for S. Note that the scattering area must also be taken into account. In general, the calculation of S is quite complex. Since 90° scatter is typically identified as the principle component of scattered radiation, the ICRP, where specific data for scattered X-ray photons is not available, allows the use of approximations that consider three different energy ranges. Below 0.5 MV, the transmission data for the primary beam may be used, i.e., the curves given in Figures 10.1, 10.2, or 10.4 may be used to calculate the shielding thickness once K is determined, because 90° scatter is assumed to have the same attenuation characteristics as the primary beam. From 0.5 MV to 3 MV the 90° scattered radiation has attenuation characteristics similar to those of 0.5 MV primary beam, so transmission data for 0.5 MV X-ray photons given in Figure 10.3 may be used. Above 3 MV, 90° scattered photons are assumed to have an energy of about 500 keV so the 1 MV transmission data for the primary beam may be used in accordance with the rule given below for gamma-ray photons. In shielding calculations for scattered radiation from gamma-ray sources, Figures 10.1 to 10.4 are used by assuming the generating potentials in kV or MV is numerically twice the photon energies in keV (MeV). For more detailed information concerning shielding from scattered radiation see the ICRP and NCRP reports [ICRP (1982 and 1991(a)), and NCRP (2004)].

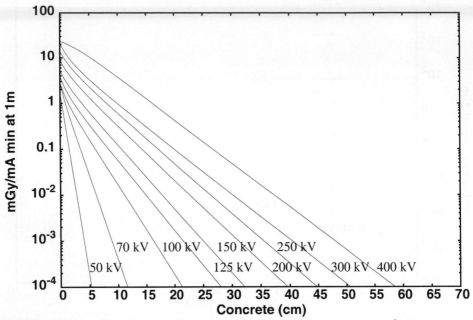

Figure 10.4: Broad–beam transmission of X-rays through concrete, density 2,250 kg· m^{-3}. 50-300 kV: half-wave generator; tungsten reflection target; total beam filtration 1 mm aluminum at 50 kV, 1.5 at 70, 2 at 100, and 3 at 125–300. 400 kV: constant potential generator; gold reflection target; 3 mm copper total beam filtration. Ordinate intercepts are 23.5 at 400 kV, 20.9 at 300, 13.9 at 250, 8.9 at 200, 5.2 at 150, 3.9 at 125, 2.8 at 100, 2.1 at 70, and 1.7 at 50. Note that the y-axis is a log scale. (Data for curves taken from ICRP Publication 33. Amsterdam, The Netherlands, Elsevier B.V., 1982.)

10.2.3 Leakage Radiation

It is necessary to also provide shielding against leakage radiation from a tube or source housing. The maximum leakage limits for different photon generators as recommended by the ICRP were given in Chapter 8. When radiation has passed through the source housing, it is appreciably attenuated and so further attenuation is virtually exponential. Hence the ICRP recommends that shielding against leakage radiation may be estimated in terms of numbers of HVL (half-value layers) or TVL (tenth-value layers). Table 10.4 contains HVL and TVL thickness of lead and concrete for various tube potentials and radionuclides. It should be kept in mind however, that if the planning technique intensity-modulated radiation therapy (IMRT) is used, the leakage radiation for linear accelerators in the range of 6 to 18 MeV is increased.

For leakage radiation, it is easiest to calculate the number of tenth-value layers N_{TVL} corresponding to the allowable transmission of a shield. To accomplish this, let us refer back to Eq. (7.7):

$$N = N_0 e^{-\mu t}$$

Table 10.3: Percent of Absorbed Dose (kerma) Rate Due to Incident Radiation Scattered to 1 m by a Tissue Like Phantom for 400 cm^2 Irradiated Area[a]

Angle of Scatter	100 kV	200 kV	300 kV	^{60}Co	6 MV
15°	–	–		– (0.48)	0.65(0.48)
30°	0.02	0.24	0.34	– (0.27)	0.30(0.24)
40°	0.03	0.23	0.26	0.18(0.14)	0.14(0.12)
60°	0.04	0.19	0.22	0.14(0.08)	0.08(0.07)
90°	0.05	0.14	0.19	0.07(0.04)	0.04(0.03)
120°	0.12	0.23	0.26	0.05(0.03)	0.03(0.02)
135°	0.17	0.30	0.33	0.04(0.02)	0.03(0.02)
150°	0.21	0.37	0.48	– (0.02)	– (0.02)

[a]Percent scattered radiation is related to primary beam measurements in air at the point of reference, i.e., at the same position as the phantom surface or phantom center. The 100, 200, and 300 kV values are from Bomford and Burlin (1963); the ^{60}Co values from Dixon, et al. (1952); the 6 MV values from Karzmark and Capone (1968); the values in parentheses from Nilsson (1975). Adapted from ICRP Publication 33, Amsterdam, The Netherlands, Elsevier B.V. 1982.

and the discussion following it. Suppose now that the original number of photons is to be reduced ten-fold. Then:

$$N = \frac{N_0}{10} \quad (10.9)$$

and:

$$\frac{N_0}{10} = N_0 e^{-\mu t_{10}} \quad (10.10)$$

Dividing both sides by N_0 and using logarithms as in Chapter 7 we obtain:

$$\ln 1 - \ln 10 = -\mu t_{10} \quad (10.11)$$

Remembering that ln 1 is always zero and also canceling the minus sign, Eq. (10.11) becomes:

$$\ln 10 = \mu t_{10} \quad (10.12)$$

and finally:

$$\frac{\ln 10}{t_{10}} = \mu \quad (10.13)$$

Using this value of μ as we did in Chapter 7, we obtain:

$$N = N_0 \, e^{-\ln 10 \cdot t/t_{10}} \qquad (10.14)$$

Using the rules for logarithms as in Chapter 7:

$$\frac{N}{N_0} = e^{-\ln 10 \cdot t/t_{10}} = 1/e^{-\ln 10 \cdot t/t_{10}} = \frac{1}{10^{t/t_{10}}} \qquad (10.15)$$

Inverting both sides of the equation gives:

$$\frac{N_0}{N} = 10^{t/t_{10}} \qquad (10.16)$$

Hence:

$$\log_{10} N_0/N = t/t_{10} = N_{TVL} \qquad (10.17)$$

In this case:

$$N_0 = \frac{W_L \cdot T}{d^2} \qquad (10.18)$$

and:

$$N = P \qquad (10.19)$$

where T and d are as usual, again U is one for leakage radiation and W_L is the weekly leakage kerma rate in air at one meter from the source. The number of half-value layers N_{HVL} is $3.3 \cdot N_{TVL}$. The necessary shield thickness can be obtained by multiplying N_{TVL} or N_{HVL} by the values given the Table 10.4.

A barrier that has been designed to shield against primary radiation is sufficient to shield against scattered and leakage radiation as well. However, if a barrier is designed for scattered and leakage radiation only, barrier thickness is computed separately for each one. If barrier thickness computed for each type of radiation differ by one TVL or more, the greater thickness is used for the barrier. If the thickness computed for each differ by less than 1 TVL the thickness of the barrier is increased by adding 1 HVL to the greater of the two computed thicknesses.

Example 10.3

A diagnostic X-ray generator is used at an average current of 30 mA for 240 minutes each week at 200 kVp. The distance between the source and the patient is 45 cm. The X-ray tube is positioned 2.0 m from a wall between the radiation room and a hematology laboratory (an uncontrolled area with infrequent air kerma). The wall contains 4.0 cm of ordinary concrete. For an occupancy factor of 1.0, what thickness of lead should by added to the wall if the primary beam is not directed to the wall? Consider that only 90° scatter and leakage radiation reach the wall.

For the scattered radiation, using Eq. (10.8), where P equals 0.1 mGy per week because the hematology laboratory is an uncontrolled area and air kerma of personnel is infrequent; d equals 2.0 m; W equals 30 mA·240 minutes per week, which equals 7200 mA minutes per week; this must be modified by $(0.45 \text{ m})^{-2}$, which equals approximately 5; T equals 1.0, and S (Table 10.3) equals 0.14, therefore, we obtain:

$$K_s = \frac{100 \cdot P \cdot d^2}{W \cdot T \cdot S} = \frac{100 \cdot \dfrac{0.1 \text{ mGy}}{\text{week}} \cdot 4 \text{ m}^2}{7200 \dfrac{\text{mA} \cdot \text{min}}{\text{week}} \cdot 5 \cdot 1 \cdot 0.14} = 7.9 \cdot 10^{-3} \frac{\text{mGy} \cdot \text{m}^2}{\text{mA} \cdot \text{min}}$$

From the 200 kVp curve in Figure 10.1 the thickness of lead required is 3.2 mm (0.32 cm) which corresponds to approximately 7.6 HVL or 2.3 TVL

For the leakage radiation, using Eq. (10.17) with Eqs. (10.18) and (10.19), where P equals 0.1 mGy per week; d equals 2.0 m, W_L equals 4 mGy per week at one meter (1 mGy per hour, for 240 minutes per week divided by 60 minutes per hour equals 4 mGy per week); and T equals 1.0, we obtain:

$$N_{TVL} = \log_{10} \frac{W_L \cdot T}{d^2 \cdot P} - \log_{10} \frac{\dfrac{4 \text{ mGy} \cdot \text{m}^2}{\text{week}} \cdot 1}{4 \text{ m}^2 \cdot \dfrac{0.1 \text{ mGy}}{\text{week}}} = 1$$

Hence one TVL is necessary to shield from the leakage radiation. Since the shielding values for scatter and leakage radiation differ by more than one TVL, the shielding for scattering (0.32 cm lead) is used. The wall already contains 4 cm of concrete which corresponds to (see Table 10.2) approximately 0.053 cm lead. Therefore, 0.267 cm of lead must be added to the wall.

10.2.3.1 Scattered Radiation from CT Units

Since a computed tomography (CT) unit continuously rotates, the amount of shielding required for these machines is determined from the scattered radiation. The American Association of Physicists in Medicine (AAPM) has, through Task Group No. 2 of their Diagnostic X-Ray Imaging committee, provided a reliable method for doing this. In its Report No. 39, the task group recommends that the unshielded air kerma to the area of interest be estimated and then compared to the occupancy-weighted permissible effective dose limit for that area. The tube potential for CT units is generally in the diagnostic range of 100 - 135 kVp. The required barrier thickness is then determined from transmission data

Table 10.4: Approximate Half-Value Layer Thicknesses and Tenth-Value Layer Thicknesses for Heavily Attenuated Broad Beams of X-ray Photons and Broad Beams of Gamma Ray Photons

	Half-Value Layer, cm		Tenth-Value Layer, cm	
X-ray Source	Lead	Concrete	Lead	Concrete
50 kV[a]	0.005	0.4	0.018	1.3
75[a]	0.015	—	0.050	—
100[a]	0.025	1.6	0.084	5.5
125[a]	—	1.9	—	6.4
150[a]	0.029	2.2	0.096	7.0
200[a]	0.042	2.6	0.14	8.6
250[a]	0.086	2.8	0.29	9.0
300[a]	0.17	3.0	0.57	10.0
400[a]	0.25	3.0	0.82	10.0
0.5 MV[a]	0.31	3.6	1.03	11.9
1[a]	0.76	4.6	2.52	15.0
2[a]	1.15	6.1	3.90	20.1
3[a]	—	6.9	—	22.6
4[a]	1.48	8.4	4.9	27.4
6[a]	1.54	10.2	5.1	33.8
10[a]	1.69	11.7	5.6	38.6
Nuclide				
positron emitters - 511 keV[b]	0.5	8.0	3.2	19.0
^{60}Co[a]	1.2	4.0	6.1	20.3
^{137}Cs[a]	0.7	2.2	4.9	16.3
^{192}Ir[a]	0.6	1.9	4.1	13.5
^{198}Au[a]	1.1	3.6	4.1	13.5
^{226}Ra	1.3	4.4	7.0	23.3

[a]Adapted from ICRP Publication 57, Elsevier, 1990.
[b]Adapted form AAPM Report 108, AAPM, College Park, MD, 2005.

numerically fit to the equation of Archer, et al. [Archer, 1983]. Transmission data for this purpose is available from D. Simpkin [Simpkin 1990]. Multi-source CT units which currently about to enter in marketplace in significant numbers, will require that the barriers take into account each source. For opposing sources, this would mean an essential doubling of the present calculations. It is necessary also to keep as mind that as CT technology in the form of multi-detector (multi-slice) and multi-source units continue to evolve, adjustments may be necessary.

10.2.4 Special Considerations for PET Units

The proliferation of positron emission tomography (PET) units usually as combined PET/CT machines, poses yet another category of shielding considerations. There is the source of an unsealed radionuclide emitting a relatively high energy (0.511 MeV) annihilation photon and the CT unit to take into account. To shield against the PET radiation only, the thickness of lead must be calculated based on the distance from the source of radionuclide which is the patient. In this circumstance, the radiation from the patient is a source of broad beam radiation which is higher in energy (0.511 MeV) than that usually encountered in diagnostic radiology. In AAPM Report No. 108, the values of broad beam transmission factors for lead, concrete, and iron given are based on consistent Monte Carlo calculations performed by one of the authors D. Simpkin [Simpkin 1989]. These were thought to be more representative of the actual situations than to consider the patient as point source. The AAPM report also contains the physical properties of the most commonly used clinical PET radionuclides and reviews typical imaging protocols and air kerma rates from patients.

The CT portion of the shielding usually can be calculated using the guidelines above, and should be calculate on the portion of time that the CT machine is in use. If it is used for ordinary diagnostic purposes as well, the calculation will need to be adjusted accordingly. This means that essentially two separate calculation need to be performed. Additionally, as the patient must generally wait for a period of 30 to 45 minutes between the injection of the radionuclide and the actually scanning session, the waiting area must be shielded as well. Again, the AAPM has studied this in great detail and produced Report No. 108 [AAPM 2005] which gives complete guidance on issue.

10.2.5 Barriers Against Neutron Generation by Linear Accelerators

As mentioned in Chapter 8, Section 8.6, the current generation of linear accelerators produce photons which have energies in the photodisintegration range (above 7 MeV), i.e., they can interact with the nucleus of the shielding material to produce neutrons. As a result, many neutrons are produced. These neutrons can contribute to the air kerma of both the patient inside as well as to persons outside the immediate facility and must be limited as much as possible. One way this is done is to chose the proper shielding material. If metal is used to provide extra shielding it must be properly chosen and placed to avoid unwanted air kerma - a metal shield on the inside of the facility may add to the patient air kerma. As an example of material choice, lead has a greater propensity to photodisintgration than does steel, so steel might be the better choice of shielding material. Steel has the disadvantage that it is more expensive, however, because of its toxicity, lead must be encased in concrete or protected by heavy coats of paint or drywall. A layer of paraffin or polyethylene, both of which are rich in hydrogen (recall hydrogen is a good absorbing material for neutrons) should be used after the primary shielding material is in place. Paraffin is less expensive than polyethylene, but has the disadvantages of being less dense and flammable. Earth and wood can also be used to shield linear accelerators. Room design, such as a maze configuration can lessen the amount of radiation that exits the facility. For a detailed discussion of shielding from X-ray photon beams, see McGinley [McGinley 1998] and NCRP reports No. 147 and 151 [NCRP 2004 and 2005].

10.2.6 Special Recommendations for Dental and Veterinary Units

Lastly when dental or veterinary units are being used to generate X-ray photons, the principles describe above for primary, and secondary barriers, and leakage radiation apply with the exceptions noted. For a detailed discussion of protection in the dental field see NCRP report No. 145 [NCRP 2003b] and for veterinary medicine see NCRP report No. 148 [NCRP 2004].

10.3 Barriers for Sealed-Sources

High activity sealed γ–ray photon sources are used mainly in radiation oncology in two forms. The first is in the form of small "seeds" which are implanted into the patient for the purpose of close irradiation of the tumor. This form of treatment is known as brachytherapy. These sources, when not in the patient, are stored in protective barriers similar to those described in Section 10.1 for unsealed sources of radionuclides. Certain precaution must be taken by the personnel when implanting such devices in the patient and the patient must be given instructions as is done for therapeutic doses of unsealed radionuclides. The only difference is that unless the sealed source leaks (a rare event) or falls out of the patient (also a rare event) there is no contamination from such sources and they are frequently permanently implanted in the patient. For a more complete discussion of protection with respect to brachytherapy source procedures see AAPM report No. 59 [NCRP 1978c].

On the other hand, radiation oncology uses apparati which contained sealed sources. These apparati may have a single sealed source (such as teletherapy units) or multiple sealed sources (such as units used for stereotactic radiosurgery, e.g., the Gamma-Knife unit). When it is necessary to calculate the thickness of barriers for a sealed-source therapy unit, Eq. (10.7) is used once more. In this case, K is replaced by the transmission, or attenuation, factor B. In addition, the work load W is expressed in units of hours per week, and the expression for B is divided by the actual air kerma rate X_r in units of mGy per hour at a distance of one meter due to the radioactivity present in the sealed source. Therefore:

$$B = \frac{P \cdot d^2}{X_r WUT} \qquad (10.20)$$

X_r can be evaluated from Eq. (10.1) as:

$$X_r = \frac{\Gamma A}{d^2}$$

where Γ is the specific gamma-ray constant for the sealed source. Since this includes only primary radiation from the source, it is generally more practical to use the measured values of air kerma because scattered radiation may make a considerable contribution for large field sizes. The thickness of the barrier required to provide an attenuation factor B is obtained from graphs prepared from experimental data by the ICRP and the NCRP. Two representative graphs are shown in Figure 10.5 for those radionuclides most commonly used as sealed sources. (Again, these are for illustration only, and primary sources should be consulted for professional work.)

Example 10.4

What thickness of primary barrier is required for a ^{60}Co teletherapy unit that provides an air kerma rate of 13 Gy per hour at a distance of 1.0 meter from the source? Assume that the use and occupancy factors are 1.0 and that the work load is 20 hours per week. The room behind the barrier is 2.0 m from the source and is a controlled area, where P equals 1 mGy per week, d equals 2.0 m, X_r equals 13 Gy·m^2 per hour or 13·10^3 mGy·m^2 per hour, W equals 20 hours per week, and U equals T equals 1. Then using Eq. (10.20) the thickness of lead required is 20.0 cm, as can be seen from Figure 10.5:

$$B = \frac{P \cdot d^2}{X_r \, WUT} = \frac{\dfrac{1 \text{ mGy}}{\text{week}} \cdot 4 \text{ m}^2}{13 \cdot 10^3 \dfrac{\text{mGy} \cdot \text{m}^2}{\text{h}} \cdot \dfrac{20 \text{ h}}{\text{week}} \cdot 1 \cdot 1} = 1.5 \cdot 10^{-5}$$

10.4 Collimation

Beam width, i.e., the cross-sectional beam area, is an important consideration whenever an X-ray or sealed-source teletherapy machine is used. The width of a photon beam is a function of focal-spot size in the case of an X-ray tube, or of the actual size of the sealed source. It is further dependent upon the cross sectional area of the opening in the shielding surrounding the primary source, i.e., on the construction of the head assembly. These two features are fixed by the manufacturer and hence cannot be altered. However, one method of externally controlling beam width is by the use of collimation.

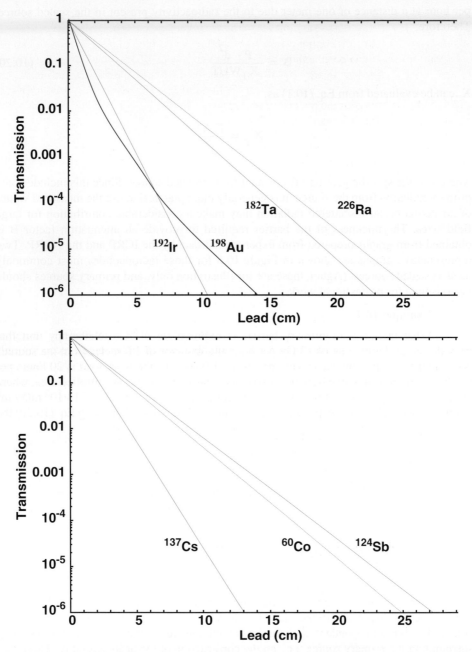

Figure 10.5: Broad-beam transmission of gamma rays from various radionuclides through lead, density 11,350 kg·m^{-3}. Note that the y-axis is a log scale. (Data for curves taken from ICRP Publication 33. Amsterdam, The Netherlands, Elsevier B.V., 1982.)

Modern collimators are usually a diaphragm system consisting of two movable pairs of metal sheets arranged at right angles to each other so that any size rectangular beam from zero to a certain maximum can be produced. In order to obtain confirmation that the requisite field size has been achieved, small mirrors are positioned inside the collimator, which reflect light from nearby light bulbs. The location of these mirrors and bulbs are adjusted so that the light emerges from the collimator in exactly the same manner as the radiation beam, when turned on, emerges from the collimator. This localizer can be used to position patients for diagnostic or oncologic procedures.

An older method of collimation was to have different fixed size cones for each field size. This collimator type usually consists of a flat sheet of metal with an attached conical metal tube extending toward the patient. A different cone must be attached to achieve each field size. The constant need to change the field size and, thus, the collimator, limits the usefulness of this type of collimator.

It must also be noted that numerous electrons are produced by the interaction of photons with the collimator. In order to prevent these electrons from reaching the patient, the collimator position should be at least 15 centimeters above the patient's skin. If the photon beam contains a large number of electrons, the energy absorbed by the skin is greatly increased and may result in severe skin reactions.

The finite size of the focal spot or sealed source causes two areas of the emergent beam to be of interest: the umbra, or region receiving the direct beam or the full strength of primary X- or gamma-ray photons, and the penumbra, which is the region receiving only some photons. A collimator is used to make the umbra a desired size for the purpose of controlling the radiation dose to the patient. At the same time, the position of the collimator between the patient and the source is chosen to lessen the penumbra as much as possible, always taking into account the production of electrons cited above.

The geometric penumbra, caused by the finite source size (focal spot or sealed source), creates an indistinct border for the radiation field. Figure 10.6 shows the umbra and penumbra on the skin surface and at a depth d within the patient. The width W of the penumbra on the patient's skin is given by:

$$W = c \left(\frac{SSD\text{-}SCD}{SCD} \right) \tag{10.21}$$

In this equation, c represents the diameter of the source; SSD is the source–skin distance; and SCD is the source–collimator distance. The width of the penumbra is independent of the radiation field size.

One method of eliminating the penumbra is to place the collimator on the skin. In this case, SSD and SCD become equal, and W is reduced to zero. This practice is undesirable because of the problem with electrons being produced by the interaction of the collimator with the primary beam, mentioned previously. Sealed sources with high specific activity and small dimensions, such as ^{60}Co, furnish a radiation beam of reasonable intensity and relatively small geometric penumbra.

The width W' of the penumbra at a depth d within a patient is given by:

$$W' = c\left(\frac{SSD + d - SCD}{SCD}\right) \qquad (10.22)$$

This shows that the penumbra at any depth within the patient is larger than that on the skin surface.

It is common to define the total penumbra on the surface or at some depth below the surface as the distance between the 10 and 90% decrement (isocontour) lines along the beam edge. These decrement lines are lines through points where the absorbed dose is a certain percent (e.g., 90, 80, 70) of the absorbed dose at the same depth along the central axis of the radiation beam. The total penumbra on and below the skin should be taken into account when diagnostic or oncologic procedures are being planned.

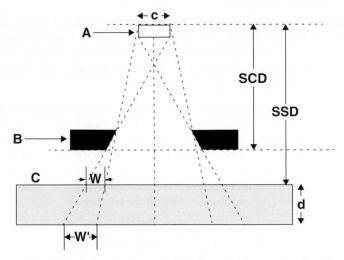

Figure 10.6: Diagram of geometric penumbra caused by finite size (denoted by **c**) of the photon source **A**. The collimator is located at **B**. The entrance side of the patient's skin is at **C**. The depth in the patient is denoted by d. The width **W** of the penumbra at **C** has widened to **W'** at a depth of **d**.

10.5 Review Questions

1.   When shielding from external sources of radionuclides, on what quantity is the minimum thickness of the shield based?

2.   What is the present acceptable standard for calculating shielding values?

3.   Using the effective/equivalent dose limits, what is the allowable effective dose for radiation workers per week? for an individual member of the general public?

4.   Briefly define each type of protective barrier.

5. Define the specific gamma-ray constant. How is it used in designing shields for small sources of radionuclides?

6. With respect to radionuclides decaying with time, how is the air kerma calculated from the air kerma rate for a given time period?

7. Define the meaning of weekly work load, W; of use factor, U; of occupancy factor, T.

8. With respect to the primary beam direction, what recommendations has the ICRP made with respect to the values of U and T for different types of areas? What differences are there for dental installations?

9. How is the difference in peak or constant kilovolts between X-ray generators taken into account? What recommendations does the ICRP make regarding the distinction between them?

10. What is the value of U for scattered and leakage radiation?

11. What is the most common object in X-ray rooms that causes scattered radiation?

12. What is the most common photon interaction with matter that causes scattered radiation below about one MeV photon energy? above one MeV photon energy?

13. Is it important to consider the angle of scatter when calculating barriers for scattered radiation?

14. What recommendations has the ICRP made regarding the calculation of scatter radiation?

15. How does the ICRP recommend that leakage radiation be calculated?

16. In terms of leakage radiation, what is the meaning of W_L?

17. Is it necessary to shield from scatter and leakage radiation if a shield for the primary beam is already in place?

18. If shielding is to be done only for secondary radiation, how is the shield for scattered radiation combined with the shield for leakage radiation?

19. How is the shielding calculated for a sealed source teletherapy unit?

20. On what does the width of a photon beam depend?

21. How is the width of the photon beam externally controlled?

22. Describe the use and function of that type of collimator known as a diaphragm system. Why is the light system both useful and necessary?

23. What are the advantages and disadvantages of fixed size cone collimators?

24. What precautions must be taken in collimator use to protect the patient? Why?

25. Define what is meant by the umbra and penumbra with regard to a broad beam X-ray photon source?

26. How is the width of the geometric penumbra defined? Why is it desirable to keep this width as small as possible?

10.6 Problems

1. The department supervisor occupies a desk that is 10 m from a 400 GBq source of ^{99m}Tc which has a physical half-life of approximately six hours. How much shielding must be placed about the source to reduce the supervisor's air kerma to 1 µGy for a period of one hour? (Assume the activity of the radionuclide remains constant over this period of time.) Consider the supervisor to be a radiation worker.

2. An X-ray machine is used at an average of 15 mA for 50 minutes each week at 200 kVp. The X-ray tube is positioned 8 m from and pointed toward a wall between the radiation room and the patient waiting area, which is fully occupied by a receptionist (assume the receptionist is a frequently exposed, non-radiation worker). The wall contains 7.5 cm of ordinary concrete. What thickness of lead must be added to the wall?

3. An X-ray generator is used at an average of 20 mA for 960 minutes each week at 300 kVp. The X-ray tube is positioned 5 m from and pointed away from the radiologist's office which he/she occupies for 20 hours each week. The wall contains 5.6 cm of ordinary concrete. What thickness of lead should be added to the wall to protect from scatter and leakage radiation? (Consider 90° scattering; assume the source to scatterer distance is 50 cm.)

4. What thickness of primary barrier is required for a ^{60}Co teletherapy unit that produces an air kerma rate of 20 Gy per hour at a distance of one meter from the source? Assume that the use and occupancy factors are one and that the work load is 40 hours per week. The room to be shielded is in a controlled area and is 4 m from the source.

Chapter 11

INTERNAL DOSIMETRY

Unsealed sources of radionuclides are generally administered internally for diagnostic or therapeutic reasons. Radiation workers, on the other hand, may unintentionally acquire an intake of such radionuclides. In either case, it is important to calculate the mean absorbed dose received not only by the total body but also by individual organs and tissues. In order to accomplish this goal, the Society of Nuclear Medicine has established the Medical Internal Radiation Dose (MIRD) committee and charged it with developing standard methods for this calculation. This committee's results are published in pamphlets issued originally as supplements to the *Journal of Nuclear Medicine* and now as regular articles in that journal. The method detailed here for calculating absorbed dose from internally deposited radionuclides uses the latest work of the MIRD Committee. As mentioned in Chapter 8, the seminal work in calculating absorbed dose from internally deposited radionuclides was of course, done by the International Commission on Radiological Protection as described in ICRP Publication 2 [ICRP, 1959].

11.1 Mean Absorbed Dose

The MIRD committee formulated the mean absorbed dose $\overline{D}(r_k \leftarrow r_h)$ to a target organ or tissue r_k from a radionuclide uniformly distributed in a source organ or tissue r_h as:

$$\overline{D}(r_k \leftarrow r_h) = \frac{\tilde{A}_h}{m_k} \sum_i \Delta_i \phi_i (r_k \leftarrow r_h) \tag{11.1}$$

$\tilde{A}_h$ the activity in the source organ or tissue accumulated over time (how this value is determined will be discussed in Section 11.3 below);

m_k the mass of the target organ or tissue r_k, (the target organ or tissue is the one for which the mean absorbed dose is to be calculated);

Δ_i the energy available from a particular decay product of the radionuclide which is distributed in the source organ or tissue r_h (this quantity is known as the equilibrium dose constant);

ϕ_i fraction of that available energy which is actually deposited or absorbed in the target organ or tissue (this quantity is known as the absorbed fraction).

The MIRD committee has used the abundant results of experimental nuclear physics to determine both the specific decay products and the frequency with which each decay product is produced for the radionuclides of interest in medical work. These results are published in MIRD Pamphlet 10 [Dillman, 1975], along with the method of obtaining the input and output data. Figure 11.1 shows the decay scheme for ^{123}I. Δ_i is obtained by multiplying the energy for *particle$_i$* (shown in the column headed E$_i$ in the output data) by the number of those particles produced per nuclear transmutation (shown in the column headed n$_i$). Δ_i is tabulated (in the old system of units) in the last column and is obtained (in the new and old system of units) in the following manner

$$\Delta_i = E_i \, MeV \cdot \frac{n_i}{disintegrations}$$

$$\Delta_{i, new} = 1.6 \cdot 10^{-13} \frac{J}{MeV} \cdot 10^3 \frac{disintegrations}{s \cdot kBq}$$

$$\times 3.6 \cdot 10^3 \frac{s}{h} \cdot E_i \, MeV \cdot \frac{n_i}{disintegrations}$$

$$= 5.76 \cdot 10^{-7} \cdot E_i \cdot n_i \frac{kg \cdot Gy}{kBq \cdot h}$$

$$\Delta_{i, old} = 1.6 \cdot 10^{-13} \frac{J}{MeV} \cdot 10^7 \frac{erg}{J} \cdot 37 \cdot 10^3 \frac{disintegrations}{s \cdot \mu Ci}$$

$$\times 3.6 \cdot 10^3 \frac{s}{h} E_i \, MeV \cdot \frac{n_i}{disintegrations}$$

$$= 2.13 \cdot E_i \cdot n_i \frac{g \cdot rad}{\mu Ci \cdot h}$$

(11.2)

When a radionuclide in some pharmaceutical form is introduced into the human body, the MIRD formulation assumes it to be distributed uniformly either throughout the entire body or throughout one or more entire organs or tissues. For this reason, the radioactive source cannot be considered as a point source. In order, therefore, to simplify the calculation, the MIRD Committee has adopted from ICRP [ICRP, 1975] a geometrically symmetric version of the human body and of human organs and tissues based on the norm that this "reference man" weighs 70 kilograms. Figure 11.2 shows the current adult human model (also known as a "phantom"). Although the tabulated values commonly available for the absorbed fraction are based on the reference man model, the MIRD committee has actually calculated them for a new born infant, a one year old, a five year old and a fifteen year old human model (which can also be used for women). As part of the reference man model, the shape and size of selected organs and tissues had to be specified.

Table 11.1 gives the mass values assigned to the organs and tissues for the 70 kg reference man model. Absorbed fraction calculations describe the amount of available energy actually absorbed by the target organ(s) or tissue(s) and thus are influenced by the geometric model used. Values for ϕ_i are only available for the tabulated organs and tissues even when the alternate human models (with the shape and size of each organ or tissue

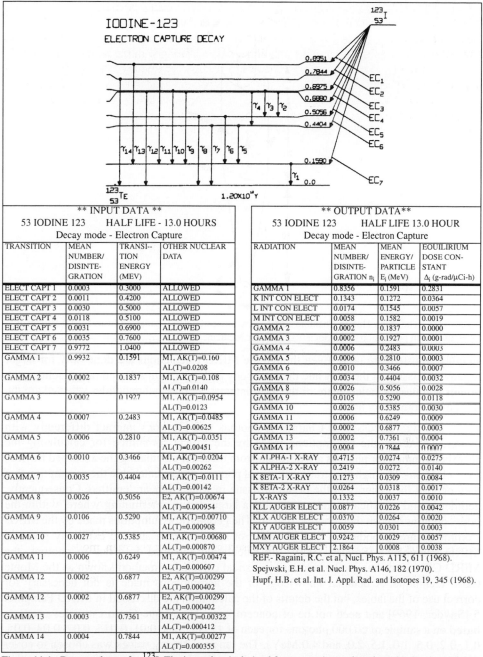

IODINE-123
ELECTRON CAPTURE DECAY

** INPUT DATA **

53 IODINE 123   HALF LIFE - 13.0 HOURS

Decay mode - Electron Capture

TRANSITION	MEAN NUMBER/ DISINTE- GRATION	TRANSI- TION ENERGY (MEV)	OTHER NUCLEAR DATA
ELECT CAPT 1	0.0003	0.3000	ALLOWED
ELECT CAPT 2	0.0011	0.4200	ALLOWED
ELECT CAPT 3	0.0030	0.5000	ALLOWED
ELECT CAPT 4	0.0118	0.5100	ALLOWED
ELECT CAPT 5	0.0031	0.6900	ALLOWED
ELECT CAPT 6	0.0035	0.7600	ALLOWED
ELECT CAPT 7	0.9772	1.0400	ALLOWED
GAMMA 1	0.9932	0.1591	M1, AK(T)=0.160 AL(T)=0.0208
GAMMA 2	0.0002	0.1837	M1, AK(T)=0.108 AL(T)=0.0140
GAMMA 3	0.0002	0.1927	M1, AK(T)=0.0954 AL(T)=0.0123
GAMMA 4	0.0007	0.2483	M1, AK(T)=0.0485 AL(T)=0.00625
GAMMA 5	0.0006	0.2810	M1, AK(T)=0.0351 AL(T)=0.00451
GAMMA 6	0.0010	0.3466	M1, AK(T)=0.0204 AL(T)=0.00262
GAMMA 7	0.0035	0.4404	M1, AK(T)=0.0111 AL(T)=0.00142
GAMMA 8	0.0026	0.5056	E2, AK(T)=0.00674 AL(T)=0.000954
GAMMA 9	0.0106	0.5290	M1, AK(T)=0.00710 AL(T)=0.000908
GAMMA 10	0.0027	0.5385	M1, AK(T)=0.00680 AL(T)=0.000870
GAMMA 11	0.0006	0.6249	M1, AK(T)=0.00474 AL(T)=0.000607
GAMMA 12	0.0002	0.6877	E2, AK(T)=0.00299 AL(T)=0.000402
GAMMA 12	0.0002	0.6877	E2, AK(T)=0.00299 AL(T)=0.000402
GAMMA 13	0.0003	0.7361	M1, AK(T)=0.00322 AL(T)=0.000412
GAMMA 14	0.0004	0.7844	M1, AK(T)=0.00277 AL(T)=0.000355

** OUTPUT DATA **

53 IODINE 123   HALF LIFE 13.0 HOUR

Decay mode - Electron Capture

RADIATION	MEAN NUMBER/ DISINTE- GRATION n_i	MEAN ENERGY/ PARTICLE E_i (MeV)	EQUILIRIUM DOSE CON- STANT Δ_i (g-rad/μCi-h)
GAMMA 1	0.8356	0.1591	0.2831
K INT CON ELECT	0.1343	0.1272	0.0364
L INT CON ELECT	0.0174	0.1545	0.0057
M INT CON ELECT	0.0058	0.1582	0.0019
GAMMA 2	0.0002	0.1837	0.0000
GAMMA 3	0.0002	0.1927	0.0001
GAMMA 4	0.0006	0.2483	0.0003
GAMMA 5	0.0006	0.2810	0.0003
GAMMA 6	0.0010	0.3466	0.0007
GAMMA 7	0.0034	0.4404	0.0032
GAMMA 8	0.0026	0.5056	0.0028
GAMMA 9	0.0105	0.5290	0.0118
GAMMA 10	0.0026	0.5385	0.0030
GAMMA 11	0.0006	0.6249	0.0009
GAMMA 12	0.0002	0.6877	0.0003
GAMMA 13	0.0002	0.7361	0.0004
GAMMA 14	0.0004	0.7844	0.0007
K ALPHA-1 X-RAY	0.4715	0.0274	0.0275
K ALPHA-2 X-RAY	0.2419	0.0272	0.0140
K 8ETA-1 X-RAY	0.1273	0.0309	0.0084
K 8ETA-2 X-RAY	0.0264	0.0318	0.0017
L X-RAYS	0.1332	0.0037	0.0010
KLL AUGER ELECT	0.0877	0.0226	0.0042
KLX AUGER ELECT	0.0370	0.0264	0.0020
KLY AUGER ELECT	0.0059	0.0301	0.0003
LMM AUGER ELECT	0.9242	0.0029	0.0057
MXY AUGER ELECT	2.1864	0.0008	0.0038

REF.- Ragaini, R.C. et al, Nucl. Phys. A115, 611 (1968).
Spejwski, E.H. et al. Nucl. Phys. A146, 182 (1970).
Hupf, H.B. et al. Int. J. Appl. Rad. and Isotopes 19, 345 (1968).

Figure 11.1: Decay scheme for [123]I. The input data is derived from experimental nuclear physics data.(From MIRD Pamphlet 10, 1975. Reproduced with permission of the Society of Nuclear Medicine.)

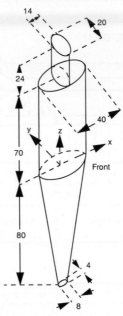

Figure 11.2: The Adult Human Phantom (dimensions in centimeters). (Adapted from MIRD Pamphlet 5, 1969, used with permission of the Society of Nuclear Medicine.)

suitably adjusted) are used. Even though the MIRD calculations are done only for larger organs and tissues of the body and the total body and for certain selected radionuclides, the organs and tissues included are of the most interest, and the radionuclides selected are the ones most commonly used.

As was discussed in Chapter 1, photons and electrons interact differently with matter. Therefore, the assumptions made by the MIRD committee for calculating the absorbed fraction for photons are different from those for calculating the absorbed fraction from electrons. The high penetrating ability of photons makes untenable the assumption that they deposit their energy in the organ or tissue in which they are emitted. Consequently, geometric and statistical assumptions play a large part in determining where, within the body, photons deposit their energy, if at all. The geometric assumptions made by the MIRD Committee are embodied in their estimate of standard man and are explicitly given in MIRD Pamphlet 5 [Snyder, 1969]. For calculating absorbed dose due to photons, the MIRD Committee has used a standard statistical calculation method known as the Monte Carlo technique. A detailed knowledge of the assumptions made is not necessary for correct use of the tables, but the details of the calculation are discussed in MIRD Pamphlet 5 [Snyder, 1969] and need not be of concern here. The Monte Carlo estimates used are based on a sample of 60,000 photons for each of 12 energies (0.01, 0.015, 0.02, 0.03, 0.05, 0.1, 0.2, 0.5, 1.0, 1.5, 2.0, and 4.0 MeV). The range of energy values was chosen to cover the range of photon values encountered in the medical use of radionuclides. Even with a sample size of 60,000 photons, there were many target organs and tissues for which the coefficient of variation (100 σ_ϕ/ϕ) of the absorbed fraction exceeded the 50% cut-off limit. For these cases the MIRD committee estimated the values by using a "buildup factor", as described in MIRD Pamphlet 2 [Berger,1968].

For electrons, there are several issues that must be considered in order to assign the value for the absorbed fraction. When all dimensions of the source organ or tissue greatly exceed the range of the electron (electron range was discussed in Chapter 1), then the absorbed fraction is given the value 1 when the source and target organ or tissue coincide and 0 otherwise. However, in the case of organs or tissues with walls which are the target for a source which is the contents, the value of ϕ is taken to be $\frac{1}{2} m_h$ where m_h is the mass of the source organ, in this case the contents. This corresponds to assuming that the absorbed fraction at the surface of the wall is one-half the absorbed fraction to the contents far from the wall. Another exception is the absorbed fraction to active bone marrow from an electron emitter deposited in bone.

The MIRD committee uses the dosimetry model of Spiers [Spiers, 1968] which implicitly assumes that cortical and trabecular bones are each uniformly labeled by the radionuclide. The absorbed fraction to the red bone marrow from trabecular bone is much greater than that from cortical bone, so if the amount of radionuclide in the two can be separated, a more accurate estimate can be made. For its calculations, however, the MIRD committee assumed that the radioactivity was equally deposited in both types of bone, which may lead to an underestimation of absorbed fraction to red bone marrow. These assumptions are specified in MIRD Pamphlet 11 [Synder, 1975], but again detailed knowledge of them is not necessary.

It is important to remember that the MIRD formulation is based on the assumption that the radioactive source is uniformly distributed in the source organ(s) or tissue(s). Other assumptions are also made which influence the accuracy of the results, e.g., when the absorbed fraction to the kidney from a source in the liver is calculated, only the average absorbed fraction to the kidney is given, whereas the absorbed fraction to the right kidney which lies nearer the liver is greater than the absorbed fraction to the left kidney. The weights and exact locations of organs and tissues in real people as opposed to the uniform 70 kg model used, may of course, be different. This effects particularly the photon absorbed fraction calculations.

Special mention should be made of the total body as a target organ when the radioactive source is located in only a few organs or tissues. In this case the source is very nonuniform with respect to the total body and the calculation of the absorbed fraction to the total body may not be very meaningful. For the 70 kg man model, absorbed fractions for specific organs and tissues for specific radionuclides are tabulated in MIRD Pamphlets 5 [Snyder 1969] and 12 [Snyder, 1977].

Table 11.1: Masses of Body Organs Used in the Reference Man Report[a]

Organ or Tissue		Organ or Tissue Mass (kg)
Adrenals		0.014
Bladder		
	Wall	0.045
	Contents	0.200
Gastrointestinal tract		
	Stomach — Wall	0.150
	Stomach — Contents	0.250
	Small intestine — Wall	0.640
	Small intestine — Contents	0.400
	Upper large intestine — Wall	0.210
	Upper large intestine — Contents	0.220
	Lower large intestine — Wall	0.160
	Lower large intestine — Contents	0.135
Kidneys (both)		0.310
Liver		1.800
Lungs (both, including blood)		1.000
Other tissues	Total	48.000
	Muscle	2.800
	Separable adipose tissue	12.500
Ovaries (both)		0.011
Pancreas		0.100
Salivary glands		0.085
Skeleton	Total	10.000
	Cortical bone	4.000
	Trabecular bone	1.000
	Red bone marrow	1.500
	Yellow bone marrow	1.500
	Cartilage	1.100
	Other constituents	0.900
Spleen		0.180
Testes		0.035
Thyroid		0.020
Uterus		0.080
Total Body		70.000

[a]Adapted from MIRD Pamphlet 11, 1975. Reproduced with permission of the Society of Nuclear Medicine.

Two other derived quantities often used in absorbed dose calculations should also be mentioned. The first is the specific absorbed fraction, Φ_i, which is defined as:

$$\Phi_i = \frac{\phi_i}{m_k} \tag{11.3}$$

where m_k is the mass of the target organ.

The second quantity is the absorbed dose per unit cumulated activity, S, which is defined as:

$$S(r_k \leftarrow r_h) = \sum_i \Delta_i \Phi_i (r_k \leftarrow r_h) \tag{11.4}$$

where Δ_i is the equilibrium dose constant defined previously and Φ_i is the specific absorbed fraction defined above.

Recall that the absorbed dose tables are based on the assumption that the radionuclide is distributed uniformly within the source organ(s) or tissue(s) in question or throughout the total body. The absorbed dose is calculated as the mean dose to the complete target organ. If a more detailed spatial distribution of dose is needed, then the tables of S values as presented cannot be used. The implications of all the assumptions made are described in MIRD Pamphlet 11. However, the S values given in the tables of MIRD Pamphlet 11 are quite sufficient for ordinary calculation of absorbed dose to patient and radiation-worker. MIRD Pamphlet 11, issued October, 1975, contains absorbed dose tables for many different radionuclides. Two of those tables, for ^{99m}Tc and for ^{123}I, are reprinted here. In the event that a particular absorbed dose calculation cannot be performed using the tables of Pamphlet 11, the earlier MIRD Pamphlets 4, 5, 6, and 10 or ICRP Publications 2 and 30, must be used.

11.2 Effective Half-life

In order to complete our discussion of mean absorbed dose, it is necessary to define the accumulated activity. To accomplish this we must first discuss the concept of effective half-life. In Chapter 1 radioactivity was defined as the rate of decay of a radionuclide. This decay rate may be expressed in the following manner:

$$A_t = A_0 e^{-\lambda t} \tag{11.5}$$

where A_t is the activity at time t, A_0 is initial activity, λ is the fractional decrease in activity per unit time, and t is time. λ is a constant, known as the decay constant, which is specific to each particular radionuclide. The resemblance between this equation and Eq. (7.7) is to be noted: the form is the same; only the names of the physical quantities have been changed (to protect the innocent of course!). Thus, this equation can follow all the same mathematical manipulation done following or with Eq. (7.7).

In Chapter 7 a half-value layer (HVL), or thickness $t_{1/2}$, was defined. The HVL was the thickness of material that reduced the initial number N_0 of photons to one-half of its original value. A similar quantity can be defined here. Since the rate of radioactive decay is directly proportional to the number of nuclei of the particular radionuclide present, the activity of the radionuclide decreases as time passes and more nuclei have decayed. When one-half of the nuclei in the sample have decayed, the rate of decay is also equal to one-half of its original value. The time for half the nuclei to decay is known as the half-life of that particular radionuclide:

$$A_t = \frac{A_0}{2} \qquad \text{when} \qquad t = t_{1/2}$$

$$(11.6)$$

and from Eq. (11.5), cancelling A_0, we obtain:

$$\frac{1}{2} = e^{-\lambda t_{1/2}} \tag{11.7}$$

Taking the natural logarithm of both sides, we have:

$$\ln \frac{1}{2} = \ln 1 - \ln 2 = -\ln 2 = -\lambda t_{1/2} \tag{11.8}$$

and, as we did in Chapter 7, we can rearrange Eq. (11.8) to obtain:

$$\frac{\ln 2}{\lambda} = t_{1/2} \tag{11.9}$$

This $t_{1/2}$ is known as the physical or radioactive half-life, $t_{1/2r}$, of the radionuclide itself.

When radionuclides are used in a biologic system, they can be eliminated from that system by metabolic processes as well as by physical decay. For each radiopharmaceutical form, a metabolic rate can be defined using the same parameters as those in the preceding equations, namely, a biologic decay constant λ_b and a biologic half-life $t_{1/2b}$. Unfortunately, nature is not always kind and sometimes the biologic activity cannot be expressed in terms of one simple exponential, as in Eq. (11.5). However, it can always be (by Euler's Theorem) expressed as some linear sum of exponentials. For simplicity, in this discussion it is assumed that the metabolic rate can be fully described by one value of λ_b and one $t_{1/2b}$. In general, the error made by this assumption is not large.

In order then to describe the total commitment of a radionuclide in a living system, the radioactive half-life must be combined with the biologic half-life. Using decay constants makes this easy:

$$\lambda_e = \lambda_b = \lambda_r \tag{11.10}$$

where λ_e is the total fractional decrease in activity per unit time, either through metabolic processes, expressed as λ_b or through radioactive decay, expressed as λ_r. Since the half-lives are related to the decay constants through Eq. (11.9), an expression for the effective half-life, $t_{1/2e}$, can be obtained. It is:

$$ t_{1/2e} = \frac{t_{1/2r} \cdot t_{1/2b}}{t_{1/2r} + t_{1/2b}} \qquad (11.11) $$

One attractive feature of this expression is that if one half-life is much larger than the other, $t_{1/2e}$ is equal to the smaller half-life. For example, suppose in a certain situation, $t_{1/2b}$ is much larger than $t_{1/2\,r}$. Then:

$$ t_{1/2r} + t_{1/2b} \approx t_{1/2b} \qquad (11.12) $$

So Eq. (11.11) gives, using Eq. (11.12):

$$ t_{1/2e} \approx \frac{t_{1/2r} \cdot t_{1/2b}}{t_{1/2b}} \qquad (11.13) $$

and hence, canceling $t_{1/2b}$ in both numerator and denominator:

$$ t_{1/2e} \approx t_{1/2r} \qquad (11.14) $$

When calculating internal dose, $t_{1/2e}$ must be used. Therefore, it is necessary to know or to approximate the value of $t_{1/2b}$, which requires a knowledge of the physiology of the living system in question. Once $t_{1/2b}$ is decided, then $t_{1/2e}$ can be calculated from Eq. (11.11).

11.3 Accumulated Activity and Residence Time

The internal absorbed dose is dependent upon a number of factors. Chief among these factors is the activity administered to the patient as well as the radioactive and biologic half-life of the radionuclide. The type of radiation emitted, and its selective energy deposition, is accounted for in another part of the calculation through the use of S, the absorbed dose per unit cumulated activity as discussed above.

When a radionuclide is administered to a patient, it diminishes in time both because of radioactive decay of the radionuclide and because of metabolic processes occurring in the person. From Eq. (11.5) the activity at any time can be expressed as:

$$ A_t = A_0 \, e^{-\lambda t} \qquad (11.15) $$

Table 11.2 S, Absorbed Dose per unit cumulated Activity, (RAD/μCi-h)

Technetium-99m   **Half-life 6.03 Hours**

SOURCE ORGANS

Target Organs	Adrenals	Bladder Contents	Stomach Contents	Intestinal SI Contents	Tract ULI Contents	LLI Contents	Kidneys	Liver	Lungs	Other Tissue (Muscle)
Adrenals	3.1E-03	1.5E-07	2.7E-06	1.0E-06	9.1E-07	3.6E-07	1.1E-05	4.5E-06	2.7E-06	1.4E-06
Bladder Wall	1.3E-07	1.6E-04	2.7E-07	2.6E-06	2.2E-06	6.9E-06	2.8E-07	1.6E-07	3.6E-08	1.8E-06
Bone (Total)	2.0E-06	9.2E-07	9.0E-07	1.3E-06	1.1E-06	1.6E-06	1.4E-06	1.1E-06	1.5E-06	9.8E-07
GI (Stomach Wall)	2.9E-06	2.7E-07	1.3E-04	3.7E-06	3.8E-06	1.8E-06	3.6E-06	1.9E-06	1.8E-06	1.3E-06
GI (Small Intestines)	8.3E-07	3.0E-06	2.7E-06	7.8E-05	1.7E-05	9.4E-05	2.9E-06	1.6E-06	1.9E-07	1.5E-06
GI (ULI Wall)	9.3E-06	2.2E-06	3.5E-06	2.4E-05	1.3E-04	4.2E-06	2.9E-06	2.5E-06	2.2E-07	1.6E-06
GI (LLI Wall)	2.2E-07	7.4E-06	1.2E-06	7.3E-06	3.2E-06	1.9E-04	7.2E-07	2.3E-07	7.1E-08	1.7E-06
Kidneys	1.1E-05	2.6E-07	3.5E-06	3.2E-06	2.8E-06	8.6E-07	1.9E-04	3.9E-06	8.4E-07	1.3E-06
Liver	4.9E-06	1.7E-07	2.0E-06	1.8E-06	2.6E-06	2.5E-07	3.9E-06	4.6E-05	2.5E-06	1.1E-06
Lungs	2.4E-06	2.4E-08	1.7E-06	2.2E-07	2.6E-07	7.9E-08	8.5E-07	2.5E-06	5.2E-05	1.3E-06
Marrow (Red)	3.6E-06	2.2E-06	1.6E-06	4.3E-06	3.7E-06	5.1E-06	3.8E-06	1.6E-06	1.9E-06	2.0E-06
Other Tissue (Muscle)	1.4E-06	1.8E-06	1.4E-06	1.5E-06	1.5E-06	1.7E-06	1.3E-06	1.1E-06	1.3E-06	2.7E-06
Ovaries	6.1E-07	7.3E-06	5.0E-07	1.1E-05	1.2E-05	1.8E-05	1.1E-06	4.5E-07	9.4E-08	2.0E-06
Pancreas	9.0E-05	2.3E-07	1.8E-05	2.1E-06	2.3E-06	7.4E-07	6.6E-06	4.2E-06	2.6E-06	1.8E-06
Skin	5.1E-07	5.5E-07	4.4E-07	4.1E-07	4.1E-07	4.8E-07	5.3E-07	4.9E-07	5.3E-07	7.2E-07
Spleen	6.3E-06	6.6E-07	1.0E-05	1.5E-06	1.4E-06	8.0E-06	8.6E-06	9.2E-07	2.3E-06	1.4E-06
Testes	3.2E-08	4.7E-06	5.1E-08	3.1E-07	2.7E-07	1.8E-06	8.8E-08	6.2E-08	7.9E-09	1.1E-06
Thyroid	1.3E-07	2.1E-09	8.7E-08	1.5E-08	1.6E-08	5.4E-09	4.8E-08	1.5E-07	9.2E-07	1.3E-06
Uterus (Nongravid)	1.1E-06	1.6E-05	7.7E-07	9.6E-06	5.4E-06	7.1E-06	9.4E-07	3.9E-07	8.2E-08	2.3E-06
Total Body	2.2E-06	1.9E-06	1.9E-06	2.4E-06	2.2E-06	2.3E-06	2.2E-06	2.2E-06	2.0E-06	1.9E-06

Decay Data Revised-March, 1972. Reference-MIRD Pamphlet 10.   Date of issue-05-13-75

(From MIRD Pamphlet 11, 1975. Reprinted with permission of the Society of Nuclear Medicine.)

^{99m}Tc

Table 11.2 S, Absorbed Dose per unit cumulated Activity, (RAD/μCi-h)

Technetium-99m   Half-life 6.03 Hours

SOURCE ORGANS

Target Organs	Ovaries	Pancreas	Red Bone Marrow	Skeleton Cortical Bone	Trabecular Bone	Skin	Spleen	Testes	Thyroid	Total Body
Adrenals	3.3E-07	9.1E-06	2.3E-06	1.1E-06	1.1E-06	6.8E-07	6.3E-06	3.2E-08	1.3E-07	2.3E-06
Bladder Wall	7.2E-06	1.4E-07	9.9E-07	5.1E-07	5.1E-07	4.9E-07	1.2E-07	4.8E-06	2.1E-09	2.3E-06
Bone (Total)	1.5E-06	1.5E-06	4.0E-06	1.2E-05	1.0E-05	9.9E-07	1.1E-06	9.2E-07	1.0E-06	2.5E-06
GI (Stomach Wall)	8.1E-07	1.8E-05	9.5E-07	5.5E-07	5.5E-07	5.4E-07	1.0E-05	3.2E-08	4.5E-08	2.2E-06
GI (Small Intestines)	1.2E-05	1.8E-06	2.6E-06	7.3E-07	7.3E-07	4.5E-07	1.4E-06	3.6E-07	9.3E-09	2.5E-06
GI (ULI Wall)	1.1E-05	2.1E-06	2.1E-06	6.9E-07	5.9E-07	4.6E-07	1.4E-06	3.1E-07	1.1E-08	2.4E-06
GI (LLI Wall)	1.5E-05	5.7E-07	2.9E-06	1.0E-06	1.0E-06	4.8E-07	6.1E-07	2.7E-06	4.3E-09	2.3E-06
Kidneys	9.2E-07	6.6E-06	2.2E-06	8.2E-07	8.2E-07	5.7E-07	9.1E-06	4.0E-08	3.4E-08	2.2E-06
Liver	5.4E-07	4.4E-06	9.2E-07	6.6E-07	6.6E-07	5.3E-07	9.8E-07	3.1E-08	9.3E-08	2.2E-06
Lungs	6.0E-08	2.5E-06	1.2E-06	9.4E-07	9.4E-07	5.8E-07	2.3E-06	6.6E-09	9.4E-07	2.0E-06
Marrow(Red)	5.5E-06	2.8E-06	3.1E-05	4.1E-06	9.1E-06	9.5E-07	1.7E-06	7.3E-07	1.1E-06	2.9E-06
Other Tissue (Muscle)	2.0E-06	1.8E-06	1.2E-06	9.8E-07	9.8E-07	7.2E-07	1.4E-06	1.1E-06	1.3E-06	1.9E-06
Ovaries	4.2E-03	4.1E-07	3.2E-06	7.1E-07	7.1E-07	3.8E-07	4.0E-07	0.0	4.9E-09	2.4E-06
Pancreas	5.0E-07	5.8E-04	1.7E-06	8.5E-07	8.5E-07	4.4E-07	1.9E-05	5.5E-08	7.2E-08	2.4E-06
Skin	4.1E-07	4.0E-07	5.9E-07	6.5E-07	6.5E-07	1.6E-05	4.7E-07	1.4E-06	7.3E-07	1.3E-06
Spleen	4.9E-07	1.9E-05	9.2E-07	5.8E-07	5.8E-07	5.4E-07	3.3E-04	1.7E-08	1.1E-07	2.2E-06
Testes	0.0	5.5E-08	4.5E-07	6.4E-07	6.4E-07	9.1E-07	4.8E-08	1.4E-03	5.0E-10	1.7E-06
Thyroid	4.9E-09	1.2E-07	6.8E-07	7.9E-07	7.9E-07	6.9E-07	8.7E-08	5.0E-10	2.3E-03	1.5E-06
Uterus (Nongravid)	2.1E-05	5.3E-07	2.2E-06	5.7E-07	5.7E-07	40E-07	4.0E-07	0.0	4.6E-09	2.6E-06
Total Body	2.6E-06	2.6E-06	2.2E-06	2.0E-06	2.0E-06	1.3E-06	2.2E-06	1.9E-06	1.8E-06	2.0E-06

^{99m}Tc

Decay Data Revised-March, 1972. Reference-MIRD Pamphlet 10.   Date of issue-05-13-75

(From MIRD Pamphlet 11, 1975. Reprinted with permission of the Society of Nuclear Medicine.)

Table 11.3 S, Absorbed Dose per unit cumulated Activity, (RAD/μCi-h)

Iodine-123   Half-life 13.0 Hours

SOURCE ORGANS

Target Organs	Adrenals	Bladder Contents	Stomach Contents	Intestinal SI Contents	Tract ULI Contents	LLI Contents	Kidneys	Liver	Lungs	Other Tissue (Muscle)
Adrenals	5.3E-03	2.0E-07	3.1E-06	1.4E-06	1.1E-06	4.6E-07	1.7E-05	6.6E-06	3.3E-06	2.2E-06
Bladder Wall	1.5E-07	2.8E-04	3.1E-07	3.3E-06	2.6E-06	8.8E-06	3.4E-07	1.9E-06	4.9E-08	2.5E-06
Bone (Total)	2.7E-06	1.0E-06	1.1E-06	1.5E-06	1.4E-06	2.3E-06	1.8E-06	1.4E-06	2.0E-06	1.3E-06
GI (Stomach Wall)	3.6E-06	3.2E-07	2.2E-04	4.8E-06	5.3E-06	2.2E-06	4.3E-06	2.3E-06	2.5E-06	1.8E-06
GI (Small Intestines)	9.8E-07	3.6E-06	3.3E-06	1.3E-04	2.6E-05	1.4E-05	3.6E-06	2.0E-06	2.3E-07	2.1E-06
GI ((ULI Wall)	1.1E-06	2.7E-06	4.7E-06	4.1E-05	2.2E-04	6.3E-06	3.5E-06	3.2E-06	2.8E-07	2.2E-06
GI (LLI Wall)	2.6E-07	9.8E-06	1.5E-06	1.1E-05	4.4E-06	3.2E-04	9.0E-07	2.7E-07	9.6E-08	2.3E-06
Kidneys	1.7E-05	3.2E-07	4.3E-06	3.9E-06	3.5E-06	1.1E-06	3.4E-04	5.1E-06	1.0E-06	2.0E-06
Liver	6.8E-06	2.2E-07	2.5E-06	2.3E-06	3.3E-06	3.1E-07	5.1E-06	7.9E-05	3.4E-06	1.5E-06
Lungs	3.2E-06	3.0E-08	2.3E-06	2.7E-07	3.1E-07	9.5E-08	1.0E-06	3.6E-06	9.2E-09	2.0E-06
Marrow (Red)	4.8E-06	2.5E-06	1.9E-06	5.3E-06	4.6E-06	7.6E-06	4.8E-06	2.0E-06	2.5E-06	2.8E-06
Other Tissue (Muscle)	2.2E-06	2.5E-06	1.9E-06	2.1E-06	2.0E-06	2.3E-06	2.0E-06	1.5E-06	2.0E-06	4.4E-06
Ovaries	6.9E-07	9.5E-06	5.7E-07	1.5E-05	1.7E-05	3.0E-05	1.3E-06	5.2E-07	1.2E-07	2.9E-06
Pancreas	1.2E-05	2.9E-07	2.7E-05	2.4E-06	2.8E-06	8.6E-07	8.5E-06	5.4E-06	3.3E-06	2.6E-06
Skin	7.1E-07	7.1E-07	5.9E-07	5.3E-07	5.4E-07	6.3E-07	7.4E-07	6.5E-07	7.2E-07	1.1E-06
Spleen	8.7E-06	5.6E-07	1.4E-05	1.9E-06	1.7E-06	9.4E-07	1.3E-05	1.1E-06	3.2E-06	2.1E-06
Testes	4.4E-08	5.9E-06	6.2E-08	3.8E-07	3.5E-07	2.3E-06	1.2E-07	8.4E-08	1.2E-08	1.6E-06
Thyroid	1.6E-07	3.5E-09	1.2E-07	2.2E-08	2.4E-08	8.4E-09	6.5E-08	2.0E-07	1.1E-06	2.1E-06
Uterus (Nongravid)	1.6E-06	2.3E-05	9.2E-07	1.3E-05	6.4E-06	8.9E-06	1.1E-06	4.5E-07	1.0E-07	3.4E-06
Total Body	3.4E-06	2.9E-06	3.0E-06	3.7E-06	3.3E-06	3.4E-06	3.4E-06	3.5E-06	3.2E-06	2.9E-06

Decay Data Revised-March, 1972. Reference-MIRD Pamphlet 10.   Date of issue-05-13-75

(From MIRD Pamphlet 11, 1975. Reprinted with permission of the Society of Nuclear Medicine.)

123I

Table 11.3 S, Absorbed Dose per unit cumulated Activity, (RAD/µCi-h)

Iodine-123   Half-life 13.0 Hours

SOURCE ORGANS

Target Organs	Ovaries	Pancreas	Red Bone Marrow	Skeleton Cortical Bone	Trabecular Bone	Skin	Spleen	Testes	Thyroid	Total Body
Adrenals	4.1E-07	1.2E-05	3.1E-06	1.5E-05	1.5E-06	9.4E-07	8.7E-06	4.4E-08	1.6E-07	3.4E-06
Bladder Wall	9.2E-06	1.7E-07	1.1E-06	6.0E-07	6.0E-07	6.6E-07	1.5E-07	6.2E-06	3.6E-09	3.5E-06
Bone (Total)	1.9E-06	1.7E-06	7.0E-06	2.0E-05	1.8E-05	1.5E-06	1.5E-06	1.0E-06	1.2E-06	4.0E-06
GI (Stomach Wall)	9.4E-07	2.8E-05	1.2E-06	6.9E-07	6.9E-07	7.1E-07	1.4E-05	4.8E-08	5.7E-08	3.5E-06
GI (Small Intestines)	1.8E-05	2.2E-06	3.3E-06	9.3E-07	9.3E-07	5.7E-07	1.7E-06	4.4E-07	1.3E-08	3.8E-06
GI (ULI Wall)	1.8E-05	2.6E-06	2.7E-06	8.7E-07	8.7E-07	5.9E-07	1.7E-06	3.8E-07	1.3E-08	3.7E-06
GI (LLI Wall)	2.4E-05	6.6E-07	3.9E-06	1.3E-06	1.3E-06	6.0E-07	7.3E-07	3.4E-06	6.8E-09	3.5E-06
Kidneys	1.1E-06	8.5E-06	2.8E-06	1.1E-06	1.1E-06	8.2E-07	1.3E-05	5.7E-08	4.2E-08	3.3E-06
Liver	6.5E-07	5.7E-06	1.1E-06	8.3E-C7	8.3E-07	7.1E-07	1.2E-06	4.1E-08	1.2E-07	3.4E-06
Lungs	7.7E-08	3.3E-06	1.9E-06	1.2E-C6	1.2E-06	8.0E-07	3.1E-06	1.0E-08	1.1E-06	3.1E-06
Marrow(Red)	7.0E-06	3.3E-06	5.5E-05	7.8E-06	1.5E-05	1.4E-06	2.1E-06	8.1E-07	1.3E-06	4.4E-06
Other Tissue(Muscle)	2.9E-06	2.6E-06	1.6E-06	1.3E-06	1.3E-06	1.1E-06	2.1E-06	1.6E-06	2.1E-06	2.9E-06
Ovaries	7.2E-03	4.4E-07	3.6E-06	1.0E-06	1.0E-06	5.0E-07	5.7E-07	0.0	7.9E-09	3.6E-06
Pancreas	5.8E-07	1.0E-03	2.0E-06	1.1E-06	1.1E-06	6.0E-07	3.0E-05	6.9E-08	9.5E-08	3.8E-06
Skin	5.1E-07	5.0E-07	8.1E-07	9.0E-07	9.0E-07	2.7E-05	6.3E-07	2.3E-06	1.1E-06	2.0E-06
Spleen	6.0E-07	3.0E-05	1.1E-06	8.0E-07	8.0E-07	7.1E-07	5.9E-04	3.0E-08	1.3E-07	3.5E-06
Testes	0.0	7.1E-08	4.8E-07	7.6E-07	7.6E-07	1.5E-06	6.6E-08	2.5E-03	9.8E-10	2.6E-06
Thyroid	7.9E-09	1.5E-07	8.5E-07	1.0E-06	1.0E-06	1.1E-06	1.1E-07	9.9E-10	4.0E-03	2.6E-06
Uterus (Nongravid)	3.0E-05	6.6E-07	2.7E-06	7.2E-07	7.2E-07	5.0E-07	4.7E-07	0.0	7.3E-09	3.9E-06
Total Body	3.9E-06	3.9E-06	3.4E-06	3.2E-06	3.2E-06	2.1E-06	3.5E-06	2.9E-06	2.9E-06	3.1E-06

Decay Data Revised-March, 1972. Reference-MIRD Pamphlet 10.   Date of issue-05-13-75

(From MIRD Pamphlet 11, 1975. Reprinted with permission of the Society of Nuclear Medicine.)

If λ_e is now used to express the decrease from initial activity A_0 to activity at any time A_t as a result of both radioactive and physiologic decay, one obtains:

$$A_t = A_0 e^{-\lambda_e t} \tag{11.16}$$

To find the activity at every time from some initial time t_1 to some future time t_2, Eq. (11.16) must be integrated with respect to time. Thus:

$$A_{\text{accumulated over time}} = \int_{t_1}^{t_2} A_0 e^{-\lambda_e t} dt \tag{11.17}$$

If t_1 is taken to be 0 and t_2 is taken to be very long with respect to $t_{1/2e}$, then t_2 is approximately equal to infinity and the integration of Eq. (11.17) yields:

$$\tilde{A}_h = \frac{A_0}{\lambda_e} \tag{11.18}$$

This expression is used in the mean absorbed dose calculation given in Eq. (11.1). In the same manner that Eq. (11.9) was derived, it is found that:

$$\frac{\ln 2}{\lambda_{1/2e}} = t_{1/2e} \tag{11.19}$$

and, by rearranging the preceding equation:

$$\frac{1}{\lambda_e} = \frac{t_{1/2e}}{\ln 2} = 1.44\, t_{1/2e} \tag{11.20}$$

Hence:

$$\tilde{A}_h = A_0 \cdot 1.44 \cdot t_{1/2e} \tag{11.21}$$

or alternatively, using Eq. (11.11):

$$\tilde{A}_h = A_0 \cdot 1.44 \cdot \frac{t_{1/2b} \cdot t_{1/2r}}{t_{1/2b} + t_{1/2r}} \tag{11.22}$$

Because radiopharmaceuticals (the result of radionuclides being tagged to physiological substances), can have various parthways in the body, the idea of "residence time" has been developed and is defined as:

$$\tau = \frac{\tilde{A}_h}{A_0} \tag{11.23}$$

If there is more than physiological compartment (for instance, sulfur colloid goes (roughly) 75% to the liver and 25% to the spleen) then Eq. 11.23 is modified to account for each compartment:

$$\tau = 1.44 \sum_i \frac{A_i}{A_0} t_{e_i} = 1.44 \sum_i F_i t_{e_i} \tag{11.24}$$

where F_i is the fraction of the administered activity at $t = 0$ for the ith exponentiated component of the time-activity function in the organ. In determining the mean absorbed dose from internally deposited radioactivity one generally uses Eq. (11.21) or Eq. (11.22) for the accumulated activity and Eq. (11.23) or Eq. (11.24) for the residence time.

11.4 Use of the MIRD Tables

We can now modify Eq. (11.1) so that the following formula can be used:

$$D_{absorbed\ dose} = \tilde{A}_{h\ accumulated\ activity}\ S_{absorbed\ dose\ per\ unit\ cumulated\ activity} \tag{11.25}$$

Using Eq. (11.21), one obtains:

$$D = A_0 \cdot 1.44 \cdot t_{1/2\ e} \cdot S \tag{11.26}$$

In the MIRD tables the units of S given for a particular radionuclide of interest must be noted. Since S is stated in rad per microcurie-hour, A_0 must be stated in microcuries, and $t_{1/2e}$ must be stated in hours. The unit rad can simply be replaced by centigray (cGy) and to convert kBq to µCi, divide by 37.

The source organ listed in the MIRD tables for S is the organ or tissue in which the radionuclide is distributed. The source organ may be total body. The target organ referred to is the organ or tissue for which mean absorbed dose is to be calculated. The source and target organs may be the same. Table 11.2 gives the S values for ^{99m}Tc and Table 11.3 gives them for ^{123}I. For other radionuclides, the actual MIRD tables would need to be consulted.

Example 11.1:

What is the absorbed dose in cGy to kidneys if 148 MBq of ^{99m}Tc sulfur colloid is administered to the patient? Assume that sulfur colloid has a biologic half-life of 4 h in the liver and ignore activity in the spleen. From Table 11.2, $t_{1/2r}$ for ^{99m}Tc is approximately 6 h. First, the effective half-life is calculated as follows:

$$t_{1/2e} = \frac{t_{1/2r} \cdot t_{1/2b}}{t_{1/2r} + t_{1/2b}} = \frac{6\ h \cdot 4\ h}{6\ h + 4\ h} = \frac{24\ h^2}{10\ h} = 2.4\ h$$

Second the initial activity can now be expressed in µCi as:

$$A_0 = 148 \text{ MBq} = 148 \cdot 10^3 \text{ kBq}/37\frac{\text{kBq}}{\mu\text{Ci}} = 4 \cdot 10^3 \mu\text{Ci}$$

From Table 11.2, S for source organ as liver (reading horizontally to the right; liver is on the first page of the table, three columns from the right-hand edge) and target organ as kidneys (reading vertically down in the left-most column of the page; kidneys are in the eighth row from the top), is 3.9E–06 rad/µCi-h. This number is found at the intersection of the third column from the right and the eighth row from the top. The notation 3.9E–06 means $3.9 \cdot 10^{-6}$; E stands for "exponent" and E–06 is to be interpreted as 10^{-6}. Consequently:

$$D = A_0 \cdot 1.44 \cdot t_{1/2e} \cdot S$$

$$= 4 \cdot 10^3 \mu\text{Ci} \cdot 1.44 \cdot 2.4 \text{ h} \cdot 3.9 \cdot 10^{-6}\frac{\text{rad}}{\mu\text{Ci} \cdot \text{h}} \cdot 1\frac{\text{cGy}}{\text{rad}}$$

$$= 53.9 \cdot 10^{-3} \text{ cGy} \approx 5.4 \cdot 10^{-2} \text{ cGy}$$

It is important to keep in mind that in the example above, we assumed all the activity administered went to the liver. In reality, this is not the case. When sulphur colloid is the pharmaceutical used, it is found that about 25% of the activity administered goes to the spleen and the rest to the liver. Hence when calculating the mean absorbed dose to a target organ or tissue from the liver as source organ, the initial activity to be used is 0.75 A_0.

Example 11.2:

What is the absorbed dose in cGy to lungs if 14.8 MBq of ^{123}I sodium iodide is administered to the patient? Assume that sodium iodide is all localized in the thyroid gland and has a biologic half-life of 17 hours. From Table 11.3 the physical half-life of ^{123}I is 13 h. As before, the effective half-life is calculated as follows:

$$t_{1/2e} = \frac{t_{1/2r} \cdot t_{1/2b}}{t_{1/2r} + t_{1/2b}} = \frac{13 \text{ h} \cdot 17 \text{ h}}{13 \text{ h} + 17 \text{ h}} = \frac{221 \text{ h}^2}{30 \text{ h}} = 7.37 \text{ h} \approx 7.4 \text{ h}$$

Secondly the initial activity can now be expressed in µCi as:

$$A_0 = 14.8 \text{ MBq} = 14.8 \cdot 10^3 \text{ kBq} \cdot 37\frac{\text{kBq}}{\mu\text{Ci}} = 0.4 \cdot 10^3 \mu\text{Ci} = 400\mu\text{Ci}$$

From Table 11.3, S for source organ as thyroid (second page of table, second column from right-hand edge) and target organ as lungs (left-most column of page, tenth row from the top), is 1.1E–06 rad/μCi–h. Hence S = 1.1·10 $^{-6}$ rad/μCi–h. Consequently.

$$D = A_0 \cdot 1.44 \cdot t_{1/2e} \cdot S$$

$$= 400\mu Ci \cdot 1.44 \cdot 7.4\, h \cdot 1.1 \cdot 10^{-6} \frac{rad}{\mu Ci \cdot h} \cdot 1 \frac{cGy}{rad}$$

$$= 46.89 \cdot 10^{-4}\, cGy \approx 4.7 \cdot 10^{-3}\, cGy$$

These examples should enable the reader to use the MIRD tables when it is necessary to calculate the mean absorbed dose to a particular organ or tissue or to the total body from a specific internally deposited radionuclide.

11.5 Review Questions

1. What is mean absorbed dose? What is its use?

2. What is meant by the equilibrium dose constant? What does it describe?

3. What is meant by the absorbed fraction? What is its role in the mean absorbed dose calculation?

4. Describe the concept of "reference man". Why was it necessary to use this geometrically simple model?

5. Why is it not possible to assume that photons deposit all their energy in the organ or tissue in which they are emitted? Why are both geometric and statistical assumptions necessary to determine where within the body, if at all, photons deposit their energy?

6. What assumptions does the MIRD committee make regarding the energy deposition by electrons in the source organ or tissue?

7. What is meant by the rate of decay of a radionuclide?

8. What is meant by the radioactive or physical half-life? the biological or metabolic half-life? the effective half-life?

9. What is meant by accumulated activity? Why is it necessary to consider this in relation to internally deposited radionuclides?

10. What is meant by residence time? How is it related to accumulated activity?

11. Describe the use of the concept of absorbed dose per unit cumulated activity.

12. Is the MIRD calculation for mean absorbed dose from internally deposited radionuclides extremely accurate? If not, why not? If not, is it sufficient for ordinary radiation protection work? In what publications may more accurate dose calculations be found?

11.6 Problems

1. A radionuclide has a physical half-life of 8 days. If it is tagged to a pharmaceutical such that its biologic half-life is 4 days, what is the effective half-life? Suppose the biologic half-life is 40 years, what is the effective half-life?

2. A patient is given a 185 MBq injection of a pharmaceutical tagged with ^{99m}Tc. The material goes entirely to kidneys and has a biologic half-life of 3 hours. What is the mean absorbed dose in cGy to the pancreas?

3. A patient is given a 148 MBq injection of a pharmaceutical tagged with ^{99m}Tc. The material has a biologic half-life of 2 hours and goes 75% to lungs and 25% to the total body. What is the mean absorbed dose in cGy to the kidneys from the source in the lungs? What is the dose to the kidneys from the remaining 25%? What is the residence time for this material in the patient?

4. A patient is given a 370 MBq injection of a pharmaceutical tagged with ^{123}I. The material goes to the total body and has a biologic half-life of 2 minutes. What is the mean absorbed dose in cGy to the total body?

Chapter 12

ABSORBED DOSE FROM EXTERNAL PHOTONS

The accurate measurement of absorbed dose from an external photon source is fraught with many of the same difficulties encountered before with internal sources (e.g., the specification of organ geometries). Furthermore, as they pass through matter, photons are absorbed or attenuated selectively, depending upon the energy of the photons and upon the density of the attenuating material. Scattered photons, principally in the form of secondary Compton photons, present a serious complication in the calculation of absorbed dose. In addition, external X-ray sources contain a wide spectrum of different photon energies.

The BEIR VII [2006] report reiterates again, that natural background radiation accounts for 82% of all the radiation exposure. Of the remaining 18%, medical procedures (medical X-ray - 58% and nuclear medicine - 21%) account for 79% of **man-made** radiation exposure to the general population. As a result, the choice between medical procedures of equal merit is often based on the absorbed radiation dose delivered to the patient. At the same time, radiation oncology procedures involving photons depend for their effectiveness on depositing a prescribed absorbed dose to a specified internal location while at the same time minimizing that to surrounding locations.

In this chapter only the basic concepts of external-dose calculations will be elucidated. For a more complete analysis of this problem the bibliographic references should be consulted.

12.1 Kerma and Absorbed Dose in Air

In Chapter 3, it was stated that kerma and absorbed dose differ in magnitude only if the energy transferred by the uncharged ionizing particles to a material is not all absorbed by that material at the point of interaction. For those electrons produced in tissue by photons with energy in the less than 12 MeV range, kerma and absorbed dose can usually be considered equal in magnitude. Recall also that the SI unit of kerma is joule per kilogram $(J \cdot kg^{-1})$. This derived unit has the common name gray (Gy).

Exposure is directly related to that part of kerma associated with the production of ionization in air. Hence, exposure and air kerma may also be considered equal when measured in the same units. Hence the units of exposure, must be translated into the units of kerma. From Chapter 3, one exposure unit is defined as 1 C/kg(air). When this is translated into the units of absorbed dose (J/kg) using for the ionization potential of air the value 33.85 eV:

$$1\frac{C}{kg} \cdot 33.85 \frac{eV}{\text{ion pair (air)}} \cdot 1.6 \cdot 10^{-19} \frac{J}{eV} \cdot 1 \frac{\text{ion pair}}{1.6 \cdot 10^{-19}C} = 33.85 \frac{J}{kg}(\text{air})$$

Hence:

$$K_{air} C/kg = 33.85 Gy \qquad (12.1)$$

One method of determining exposure is to use a radiation measuring instrument, such as a calibrated ionization chamber, designed to approximate charged particle equilibrium conditions.

12.2 Absorbed Dose to a Small Mass of Tissue Exposed in Free Space

Suppose now that absorbed dose to the center of a small mass of soft tissue Δm situated in free space is to be determined. The mass chosen must be large enough to avoid statistical fluctuations due to photon interactions taking place in it. For a given photon beam, the energy absorbed is directly proportional to the mass energy absorption coefficient (μ/ρ, the linear attenuation coefficient divided by the density—mass per unit volume) of the stopping material. This quantity is a measure of how much of the energy transferred in a given collision is absorbed in the material or medium. Since this is difficult to determine for an individual case, an average value is generally used. Hence:

$$K_{\Delta m} = K_{air} \frac{(\bar{\mu}/\rho)_{med}}{(\bar{\mu}/\rho)_{air}} \qquad (12.2)$$

where the ratio of the average mass energy absorption coefficient for the material Δm to air is used.

To obtain the absorbed dose rather than kerma, the mass Δm must be surrounded by additional material so that it is made large enough that electron equilibrium will be established. After exposure to a beam of photons, the kerma to Δm can be determined by placing a calibrated ion chamber at the center of Δm. The size of the ion chamber is not important, but its walls must be thick enough to establish electron equilibrium. The kerma to Δm can then be calculated from:

$$K'_{\Delta m} = K_{\Delta m} \cdot A_{eq} \tag{12.3}$$

A_{eq} is a factor slightly less than 1.0, which gives the fraction of photon beam transmitted through a thickness r_{eq} of material. It must be emphasized that Eqs. 12.2 and 12.3 represent different physical situations. Eq. (12.2) describes the kerma to the mass itself, whereas Eq. (12.3) gives the kerma to the small mass Δm surrounded by enough additional material to produce electron equilibrium.

12.2.1 A_{eq} and r_{eq}

This analysis then depends upon determining the value of A_{eq}. In turn, the value of A_{eq} depends upon the thickness r_{eq} of tissue required to establish electron equilibrium. So the main difficulty in obtaining a value for A_{eq} is determining a value for r_{eq}. This latter consideration is made difficult because even a monoenergetic photon beam can set in motion electrons of all energies, from zero to some maximum value. It is standard to choose for r_{eq} the mean range of electrons produced by the photon beam. Consequently, for monoenergetic photons with energies between 200 and 400 keV, A_{eq} has a recommended value of 1.0. For high-energy photons, the value of A_{eq} decreases until it is about 0.90 for photon energies of 25 MeV. The precise value of A_{eq} cannot be determined, but experimental values have been obtained by Cunningham and Sontag [1980]. Commonly, a value of 1.00 is used for low photon energies; a value of 0.99, for intermediate energies (^{137}Cs range); and a value of 0.985 or 0.978 for the [60]Co–3 MeV Linac range. On the basis of accepted experimental data, workers in this field commonly concur that A_{eq} should be used as just stated.

To determine absorbed dose to the tissue Δm exposed in free space, one further step must be taken. For a given photon beam, the energy absorbed is directly proportional to the mass energy absorption coefficient $(\mu_{en})_m$ of the stopping material, which is the energy absorption coefficient divided by the density ρ of this material. Density is the mass per volume of material and the energy absorption coefficient is the energy actually absorbed by the material and not scattered from it. It is given by $\mu_{en} = \mu \cdot \dfrac{E_a}{E_\gamma}$, where μ is the linear attenuation coefficient, E_a is energy actually absorbed, and E_γ is the energy of the photon. The dose to the medium Δm is given by:

$$D_{med} = \left[\frac{(\mu_{en}/\rho)_{med}}{(\mu_{en}/\rho)_{air}}\right] \cdot D_{air} = \left[33.85 \cdot \frac{(\mu_{en}/\rho)_{med}}{(\mu_{en}/\rho)_{air}}\right] \cdot K_{air} \cdot A_{eq} \tag{12.4}$$

Rigorously, A_{eq} used here should be slightly different from the value used in Eq 12.3 which applies to air. It will be assumed, however, that they are approximately the same.

12.2.2 f Factor

The quantity in square brackets in Eq. (12.4) is very important in dose calculations. It is given the symbol f_{med} and is known as the "f factor":

$$f_{med} = 33.85 \cdot \frac{(\mu_{en}\rho^{-1})_{med}}{(\mu_{en}\rho^{-1})_{air}} \qquad (12.5)$$

Figure 12.1 is a plot of the f factor for monoenergetic photons. As can be seen, for water and muscle, the f factor does not vary much over the entire energy range from 0.01 to 10 MeV. This lack of variance occurs because air, water, and muscle have essentially the same effective atomic number and the same number of electrons per gram; consequently, the quantity $\mu_{en}\rho^{-1}$ is similar for these three materials. Bone, however, has a higher effective atomic number and shows a rapid drop in f for photon energies between 0.04 and 0.1 MeV, where the photoelectric effect ceases to be an important mechanism for stopping photons. Below 0.1 MeV, however, the f factor for bone is relatively high. This is of importance for calculating absorbed dose to red bone marrow.

When a material is exposed to a polyenergetic photon beam, such as a diagnostic X-ray beam, it is necessary to weight the f values obtained from Figure 12.1 by the quantity of photons in beams having each specific energy. Alternatively, the HVL for the kVp used can be employed to obtain the appropriate f factor.

12.3 The Bragg-Gray Principle

The determination of absorbed dose from a measurement of air kerma, as just described, is assuredly the most convenient way of measuring absorbed dose; however, this method is limited because it is only possible to obtain the absorbed dose in free space. The Bragg-Gray principle, which relates absorbed dose to a measurement of the actual amount of ionization produced in a small, gas-filled cavity in the material, is an attempt to surmount these difficulties.

When a solid material is traversed by a beam of photons, electron tracks are produced, as shown in Figure 12.2. If now a small, gas-filled cavity is placed in the material, ionization will be produced in the cavity as well. The Bragg-Gray principle states that, as long as the cavity is so small that it does not alter the number or distribution of electrons that pass through the material, the ionization produced in the cavity can be related to energy absorbed in the material at the position of the cavity. This principle allows determination of absorbed dose at different depths within the material as well as for different densities. It also gives an accurate method of determining experimentally absorbed dose at any particular point in a material such as tissue from any particular X-ray beam.

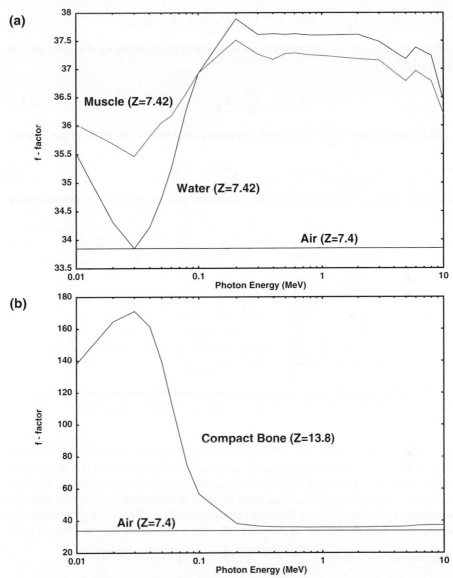

Figure 12.1: (a) The f-factor as a function of photon energy for water, muscle; (b), for bone. (Data for curves taken from *Medical Imaging Physics 4th Ed.*, W.R. Hendee and E.R. Ritenour, New York, John Wiley and Sons, 2002.)

To find the actual dose absorbed by a material, first the energy transferred to the gas in the cavity must be determined. The energy deposited per kilogram of gas in the cavity is given by:

$$E_g = J_g \frac{C}{g \, kg} \cdot W \frac{J}{C} = (J_g \cdot W) \frac{J}{kg} = J_g \cdot W \, Gy \qquad (12.6)$$

The corresponding energy E_m imparted to a unit mass of the material by the same photon beam is given by:

$$E_m = S_g^m \cdot E_g \qquad (12.7)$$

In Eq. (12.7) the quantity S_g^m is the mass stopping-power ratio of material to gas. Hence:

$$E_m = S_g^m \cdot E_g = S_g^m \cdot J_g \cdot W \qquad (12.8)$$

Each quantity in Eq. (12.8) must now be examined and its method of determination discussed.

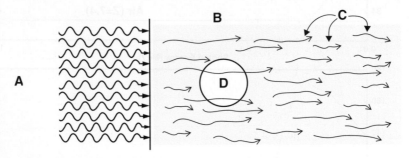

Figure 12.2: Illustration of Bragg-Gray cavity in a medium traversed by electron tracks. The photon beam, **A**, enters the material, **B**, and creates electron tracks, **C**. The small gas-filled cavity, **D** (size enlarged), does not alter the electron tracks.

The quantity J_g is determined experimentally. A cavity of known volume filled with a gas of known density and ionization potential is placed within the material. In practice, calibrated ion chambers are used. However, the discussion given here is useful. The total quantity of charge produced by the irradiating photons is measured using a very sensitive charge-measuring device, such as a Keithley electrometer. The volume of a spherical cavity can be determined by measuring its diameter (d) with a measuring device, such as a micrometer, and using the formula $\frac{4}{3}\pi\left(\frac{d}{2}\right)^3$ for the volume of a sphere. Thus:

$$J_g = \frac{\text{total charge measured}}{\text{mass}}$$
$$= \frac{\text{total charge measured}}{\text{density} \cdot \text{volume}} \qquad (12.9)$$

W is the average amount of energy (J) needed to liberate one unit charge (C). Since the energy needed to ionize material is usually stated in eV, this quantity must be changed to joule. Consequently:

$$W = \frac{\dfrac{\text{Average ionization energy eV}}{\text{ion pair}} \cdot \dfrac{1.6 \cdot 10^{-19} \text{J}}{\text{eV}}}{\dfrac{1.6 \cdot 10^{-19} \text{C}}{\text{ion pair}}} \qquad (12.10)$$

and so:

$$J_g \cdot W = \frac{\text{Total charge measured C} \cdot \text{Average energy J}}{(\text{density kg} \cdot \text{m}^{-3}) \cdot (\text{volume m}^3) \cdot \text{charge C}} = E_g \text{ Gy} \qquad (12.11)$$

The quantity S_g^m is not so easy to obtain. Many methods for obtaining it have been employed. Table 12.1 gives some mass stopping-power ratios ($\bar{S}_g^m$) relative to air for several different materials, for some photon energies. The density of air is approximately 1.3 kg·m^{-3} at standard temperature and pressure (STP: 0°C, 101.3 kPa). Most chambers are calibrated at normal temperature and pressure (NTP: 20°C, 101.3 kPa) where the density of air is 1.2 kg· m^{-3}. The absorbed dose to the material is then:

$$D_m = E_m = J_g \cdot W \cdot \bar{S}_g^m \qquad (12.12)$$

With the average energy per ion pair for air taken as 33.85 eV, Eq. (12.12) becomes (at STP).

$$D_m = E_m = \frac{\text{Total charge measured} \cdot 33.85}{1.3 \cdot (\text{volume})} \cdot \bar{S}_g^m \cdot \text{Gy} \qquad (12.13)$$

Example 12.1

A block of lucite is irradiated with photons from a ^{60}Co source. A cavity with a volume of $1.0 \cdot 10^{-6}$ m^3 is placed in the block. Calculate absorbed dose to lucite at the position of the cavity after a charge of 10^{-7} C has been measured by the electrometer. Assume the cavity to be filled with air at STP. $\bar{S}_{air}^{lucite}$ for ^{60}Co is given by Table 12.1 as 1.109. Using Eq. (12.13) we obtain:

$$D_{lucite} = \frac{10^{-7} \cdot 33.85}{1.3 \cdot 1.0 \cdot 10^{-6}} \cdot 1.109 \text{ Gy}$$

which results in:

$$D_{lucite} = 29 \cdot 10^{-1} \text{ Gy} = 2.9 \text{ Gy}$$

Hence the absorbed dose to the lucite is 2.9 Gy.

Table 12.1: Mean Mass Stopping-Power Ratios $\overline{S}\,_g^m$

Gamma-Ray Emitter	Energy of Radiation	Poly-ethyle	Water	Tissue (Muscle)	Poly-styrene	Lucite	Gra-phite
^{198}Au	0.41 MeV	1.233	1.149	1.149	1.139	1.124	1.031
^{137}Cs	0.67 MeV	1.225	1.145	1.145	1.133	1.120	1.010
^{60}Co	1.25 MeV	1.209	1.135	1.133	1.120	1.109	1.002

12.4 Other Absorbed Dose Calculation Methods

Many other methods exist for obtaining absorbed dose from external photon beams beside those described here. One can determine absorbed dose in an extended material by using calibrated ion chambers. In fact, this is the preferred method. It is also possible to measure absorbed dose by calorimetry methods. In addition, various methods of chemical dosimetry have been employed. The use of thermoluminescent devices (TLD's) and optically stimulated luminescence (OSL) to measure absorbed dose are more recent developments, and film is widely used as well. All of these methods present certain advantages and certain disadvantages. That method which is most advantageous for the particular problem at hand should be employed. Depth-dose calculations are of prime importance in radiation oncology treatment planning; skin entrance air kerma plays a similar role in choosing diagnostic X-ray procedures.

12.5 Absorbed Dose in Selected Diagnostic X-ray Procedures

The United States Center for Devices and Radiological Health/FDA (formerly the Bureau of Radiological Health) has produced both a comprehensive booklet and a small, pocket-sized pamphlet summarizing absorbed dose to selected vital organs and the total body from common diagnostic X-ray procedures. (ICRP Publication 34 [1982(a)], contains a similar set of tables and calculation.) These tables are based on the same model for reference man used in Chapter 11 for internal dosimetry. In fact, the Monte Carlo techniques and geometric considerations developed by the MIRD committee have been taken over precisely for these external dose estimates. The beam quality (kVp technique used) is stated in terms of HVL's (see Chapter 7) of aluminum. The mAs (X-ray tube current, mA, multiplied by the air kerma time in seconds, s) is stated in terms of an assumed entrance kerma of 0.876 cGy free−in−air. It is recommended that, at the minimum, anyone engaged in diagnostic radiology obtain the pamphlet and learn how to use it. One table from this pamphlet is reproduced here as Table 12.2. A sample calculation is done to explicate its use.

Table 12.2: Sample X-ray Doses[a]

Shoulder: Organ dose (Gy) for 0.873 cGy entrance kerma (free–in–air)
Condition: SID: 1.016 m (40 in)
           (source–to–image receptor distance)
        Film Size = Field Size—see each projection.
        Entrance Kerma (free–in–air): 0.873 cGy
Projection: AP Shoulder (right or left only)
        0.254 m · 0.305 m (10 by 12 in)

Dose: 10^{-5} Gy

Beam Quality HVL (mm Al)

	1.5	2.0	2.5	3.0	3.5	4.0
Testes	*	*	*	*	*	*
Ovaries	**	**	**	**	**	**
Thyroid	11	16	21	26	30	35
Active bone marrow	4.6	6.5	8.5	11	13	16
Embryo (uterus)	***	***	***	***	***	***

Projection: AP Shoulder (both)
        0.356 m · 0.432 m (14 by 17 in)

Dose: 10^{-5} Gy

Beam Quality HVL (mm Al)

	1.5	2.0	2.5	3.0	3.5	4.0
Testes	**	**	**	**	**	**
Ovaries	**	**	**	**	**	**
Thyroid	519	648	748	824	882	927
Active bone marrow	15	21	29	37	47	58
Embryo (uterus)	***	***	***	***	***	***

* No detectable contribution;   ** < 10^{-7} Gy;   *** < 10^{-8} Gy.

[a]From Rosenstein, 1976. When attempting to use the pamphlet, the instructions stated on page 3 should be noted. The beam quality (HVL, mm Al) and entrance kerma free–in–air (Gy) must be determined at the facility. This can be accomplished by direct measurement or by use of one or more of the publications cited at the end of the pamphlet. It is recommended that linear interpolation between listed HVL's be done. When the source–to–image receptor distance (SID) is within 25 centimeters (10 inches) of the listed SID, ignoring this distance difference results in dose variations no larger than 10%. Special precautions are prescribed for calculating absorbed dose to active bone marrow and to testes.

Example 12.2

Use Table 12.2 to calculate absorbed dose to thyroid using two different radiographic procedures.

	CASE A	CASE B
View, projection	AP shoulder, left and right shoulders on separately exposed radiographs	AP shoulder; both shoulders on one radiograph
Field size (at image receptor)	0.254 by 0.305 m (10 by 12 in) (each radiograph)	0.356 by 0.432 m (14 by 17 in)
SID	1.016 m (40 in)	1.016 m (40 in)
Beam quality (HVL)	2.5 mm Al	2.5 mm Al
Entrance kerma (free–in–air)	0.218 cGy	0.218 cGy
Organ of interest	Thyroid	Thyroid
(1) From Table 12.2 the thyroid absorbed dose for 0.873 cGy for HVL = 2.5 mm Al	$21 \cdot 10^{-5}$ Gy (each radiograph) = $42 \cdot 10^{-5}$ Gy (total)	$748 \cdot 10^{-5}$ Gy
(2)The thyroid absorbed dose for 0.218 cGy $(1/4 \cdot 0.873$ cGy) is therefore:	$10.5 \cdot 10^{-5}$ Gy (for standard man where the shoulders are assumed to be 31 cm from the phantom vertex)	$187 \cdot 10^{-5}$ Gy

Note the significant reduction (94%) in absorbed dose to the patient's thyroid for Case A. Case A can be achieved either by exposing two separate small detectors (electronic or film) or by simply blocking the thyroid area with an appropriate external shield during air kerma of the larger detector. The gain is related to X-ray technique alone.

12.6 Conclusions

In general, absorbed dose from most diagnostic radiographic procedures is low. However, the radiation absorbed dose from procedures involving fluoroscopy and angiography can be considerably higher. Particular care should be taken when employing these techniques to keep patient absorbed dose as low as possible. In general, the absorbed dose to the patient from diagnostic radiology procedures is not regulated. The two exceptions are fluoroscopy, where the absorbed dose rate is limited to a maximum of 0.1 Gy/min at the table top or at 30 cm from the image intensifier C-arm, and mammography, where the average glandular absorbed dose is limited to a maximum of 3 mGy per view for the standard breast defined as 4.5 cm in size consisting of 50% glandular and 50% adipose tissue. It is also recommended that a radiation worker performing these techniques takes as great advantage as possible of the three basic principles of radiation protection: time, distance, and shielding.

Finally, a word should be said about the type of absorbed dose encountered in reconstructed X-ray tomography (CT). The absorbed dose to the patient uniformly distributed throughout the slice, is high by usual standards employed for diagnostic X-ray procedures. The absorbed dose from CT depends on mAs per slice, but there is also some contribution to absorbed dose, in the form of scatter, from the nearby slices. The current generation of multi-detector (multi-slice) systems have a slightly higher absorbed dose to the patient than the former generation of machines. The United States Food and Drug Administration (FDA) uses the computed tomography dose index (CTDI) to approximate the absorbed dose for a large set of contiguous slices. This includes the absorbed dose resulting from each slice as well as the additional absorbed dose because of field overlap and scatter from adjacent slices. The absorbed dose from a large set of slices is about twice the absorbed dose to a single slice because of the effect of scatter. Most narrow slice techniques tend to increase the absorbed dose to the patient both because mAs increases for narrower slice width and because there is more field overlap for narrower slices. Additionally, the coming generation of multi-source machines will most likely increase the absorbed dose to the patient slightly, though that is yet to be seen. However, in this case as in every case, the medical benefit to the patient must be balanced against the expected radiation effects. As long as this is done, there should be no hesitancy in employing this or any other diagnostic X-ray technique.

The absorbed dose to the patient from therapeutic X-ray and sealed-source techniques is also high. Protocols, such as rotation techniques, dynamic collimation, intensity-modulated radiation therapy (IMRT), and image guided radiotherapy (IGRT) have been developed to minimize absorbed dose to those portions of the patient not undergoing treatment and to maximize absorbed dose to treatment site(s). Treatment planning is only done on an individual patient basis. As always, the potential benefit to the patient is weighed against possible harmful radiation effects.

12.7 Review Questions

1. What is the largest component of man-made radiation exposure to the general population?

2. How are kerma and absorbed dose related? How is exposure related to air kerma determined?

3. What is the meaning of A_{eq}? On what does the value of A_{eq} depend?

4. What is the meaning of r_{eq}? Why is it difficult to establish a value for it? what value is usually chosen for r_{eq}?

5. What is the "f factor"?

6. What is the Bragg-Gray principle? Why was it necessary to postulate it?

7. How is the quantity J_g defined? W? $S\frac{m}{g}$?

8. What other methods exist for obtaining absorbed dose from external photon beams?

9. What absorbed dose estimates have been provided by the Center for Devices and Radiological Health and the ICRP? On what standard model are these estimates based?

12.8 Problems

1. Using Figure 12.1, find the f factor for photons having an energy of 0.02 MeV and for those of energy 2.0 MeV, in water, muscle and bone.

2. A tissue phantom is irradiated by a sealed source of ^{137}Cs. A cavity having a volume of $2.0 \cdot 10^{-6}$ m^3 is embedded in the phantom. Calculate absorbed dose to tissue at the position of the cavity after a charge of 10^{-10} C has been measured by the electrometer. Assume the cavity is filled with air at STP.

3. A water phantom is irradiated by a sealed source of ^{198}Au. A cavity having a volume of $1.0 \cdot 10^{-6}$ m^3 is embedded in the phantom. Calculate the charge measured by the electrometer when the absorbed dose is 5 Gy. Assume the cavity is filled with air at STP.

4. Compare the absorbed dose to active bone marrow using the two different radiographic procedures for an AP of the shoulders. Assume 2.0 mm Al for the beam quality and the entrance kerma (free–in–air) is 0.5 cGy.

Appendix A

THE UNITS OF PHYSICS

Physics is an axiomatic system, just like Euclidean geometry taught to all high school students. In geometry, the concepts of point and line are not defined. It is assumed that the student has an intuitive grasp of the meaning of these terms. The concepts of point and line are known as undefined terms. In physics, the undefined terms are the concepts of mass, length, time, and current. All derived quantities, such as energy, speed, and voltage, are defined in terms of these concepts.

Three systems of units are in general use for describing physical quantities. The English system, because of its limited use, will not be discussed. The other two systems of units are both based on the metric system. The first of these is known as the cga (centimeter–gram–second) system. This system is being phased out of use. The second system is known as the International System of Units with the international abbreviation SI (from the French name, Le Systeme International d'Unites). The SI system replaces the system of units formerly known as the MKSA (meter–kilogram–second–ampere) system. Table A.1 gives a summary of the base SI Units.

Table A.1: Base SI Units

Quantity	Unit Name	Symbol
length	meter	m
mass	kilogram	kg
time	second	s
electric current	ampere	A
thermodynamic temperature	kelvin	K
luminous intensity	candela	cd
amount of substance	mole	mol

In 1790 the French National Assembly requested that the French Academy of Sciences formulate a system of units suitable for adoption by the entire world. This system was based on the meter as a unit of length and the kilogram as a unit of mass. It was adopted by commerce and industry and soon also in scientific and technical circles. On June 22, 1799 two platinum standards, one for the meter (today the standard for length is defined in terms of one wavelength of krypton-86 instead of the standard meter stick) and one for the kilogram were deposited in the Archives de la Règublique in Paris. Article 1 Section 8 of the United States Constitution, written in 1787, recognized the importance of regulating weights and measures. However, units in the metric system were not legalized in the United States until 1866 and the international meter and kilogram did not became the fundamental standards of length and mass until 1893.

International standardization began with the meeting of fifteen nations in Paris in 1870. This led to the Meter Convention (treaty) which was signed in Paris in 1875 by representatives of seventeen nations. As of 1 December 2006, there were fifty-one member states. This treaty established the General Conference on Weights and Measures (CGPM — Conference Generale des Poids et Mesures) to handle all international matters concerning the metric system. It also established the International Bureau of Weights and Measures (BIPM — Bureau International des Poids et Mesures) with headquarters in the grounds of Parc de Saint-Cloud, Sèvres near Paris. Its principal building, the Pavillon de Breteuil, was occupied in 1875. This was placed at the disposal of the BIPM by the French government and its upkeep is financed jointly by the Member States of the Meter Convention. Additionally, the Meter Convention established the International Committee for Weights and Measures (CIPM — Comité International des Poids et Mesures). The CGPM currently meets every four years in Paris (the 22nd meeting took place in October 2003). The CGPM controls the BIPM, which preserves the metric standards, compares national standards with them and conducts research to establish new standards. The CIPM consists of eighteen members and meets every year in September or October at the BIPM. It is the CIPM which submits requests for changes to the metric system to the CGPM for official international adoption. The National Institute for Science and Technology (formerly the National Bureau of Standards) represents the United States in these activities.

In 1954 the CGPM adopted a rationalized and coherent system of units to which in 1960 it gave the name International System of Units and the international abbreviation SI. There were originally six base units with a seventh (mole) added in 1971 (see Table A.1).

In 1975 the CGPM approved two special names: the gray (Gy) as a special name for the SI unit of absorbed dose—one joule per kilogram; and the becquerel (Bq) as a special name for the SI unit of activity of a radionuclide—one disintegration per second. In 1979 the special name sievert (Sv) was approved for the unit of equivalent dose—also one joule per kilogram. Table A.2 lists some of the derived SI units with special names. In 1980, SI units were adopted by the International Commission on Radiation Units and Measurements (ICRU), for use in research work and application in radiology. The ICRU has urged that SI units be used completely within as short a period as possible. By the end of 1994, SI units had been officially adopted for radiological applications by every country.

Table A.3 lists the common prefixes used with various powers of ten.

Table A.2: Derived SI Units with Special Names[a]

Quantity	Unit	Symbol	Formula
Absorbed dose, specific energy imparted, kerma, absorbed dose index	gray	Gy	J/kg
Activity (of a radionuclide)	becquerel	Bq	1/s
Dose equivalent, Ambient dose equivalent, Directional dose equivalent, Personal dose equivalent, Equivalent dose, Effective dose	sievert	Sv	J/kg
plane angle	radian	rad	$m \cdot m^{-1} = 1$
solid angle	steradian	sr	$m^2 \cdot m^{-2} = 1$
Celsius temperature	degree Celsius	°C	°K − 273.15
Catalytic activity	katal	kat	mol/s
Electric capacitance	farad	F	C/V
Electric charge, quantity of electricity	coulomb	C	A·s
Electric conductance	siemens	S	A/V
Electric inductance	henry	H	Wb/A
Electric potential difference, electromotive force	volt	V	W/A
Electrical resistance	ohm	W	V/A
Energy, work, quantity of heat	joule	J	N·m
Force	newton	N	$kg \cdot m/s^2$
Frequency (of a periodic phenomenon)	hertz	Hz	1/s
Illuminance	lux	lx	lm/m^2
Luminous flux	lumen	lm	cd·sr
Magnetic flux	weber	Wb	V·s
Magnetic flux density	tesla	T	Wb/m^2
Power, radiant flux	watt	W	J/s
Pressure, stress	pascal	Pa	N/m^2

[a] Adapted from Federal Register Vol. 63, No. 144 Tuesday, July 28, 1998, published by the Department of Commerce, National Institute of Standards and Technology.

Table A.3: Common Prefixes

Power of 10	Name	Symbol
10^{24}	yotta	Y
10^{21}	zetta	Z
10^{18}	exa	E
10^{15}	peta	P
10^{12}	tera	T
10^{9}	giga	G
10^{6}	mega	M
10^{3}	kilo	k
10^{2}	hecto	h
10^{1}	deca	da
10^{-1}	deci	d
10^{-2}	centi	c
10^{-3}	milli	m
10^{-6}	micro	μ
10^{-9}	nano	n
10^{-12}	pico	p
10^{-15}	femto	f
10^{-18}	atto	a
10^{-21}	zepto	z
10^{-24}	yocto	y

Appendix B

THE ROENTGEN AND THE RAD

B.1 The Roentgen

Initially, the unit of exposure was defined in terms of the cgs system of units and was given the name roentgen.[1] In the original definition the roentgen was that exposure of photons which ionized enough air to produce a charge of 1 esu of either sign in a 1 cc (cm^3) volume of air:

$$1\ R\ =\ 1\,esu \cdot cc^{-1}(air) \tag{B.1}$$

At STP 1 cc of air weighs 0.001293 g. Hence:

$$1\ R\ =\ \frac{1\ esu}{0.001293\ g}\ (air) \tag{B.2}$$

The equivalence between esu and coulomb is as follows:

$$3 \cdot 10^9\,csu\ =\ 1\ C \quad \text{and} \quad 1\ g = 10^{-3}\ kg \tag{B.3}$$

As a result:

$$1\ R\ =\ 2.58 \cdot 10^{-4}\ C \cdot kg^{-1} \tag{B.4}$$

B.2 The Rad

The absorbed dose has been defined in Chapter 3 as the quantity of energy imparted by any ionizing radiation to any material substance per unit mass of that substance. The rad is defined as 100 erg of energy deposited per gram mass:

$$1\,rad = 100\ erg \cdot g^{-1} \tag{B.5}$$

---

1. The name "roentgen" was given to the unit of exposure to honor Wilhelm Conrad Röentgen (1845-1923), who discovered the phenomenon of X radiation in 1895. He was awarded the Nobel prize in physics in 1901.

The term "rad" is the symbolic name for "radiation absorbed dose". The equivalence between erg and joule is as follows:

$$1\,\text{erg} = 10^{-7}\,\text{J} \tag{B.6}$$

Hence:

$$1\,\text{rad} = \frac{100\,\text{erg}}{\text{gm}} \cdot 10^{-7}\frac{\text{J}}{\text{erg}} \cdot \frac{1\,\text{g}}{10^{-3}\,\text{kg}} = 10^{-2}\frac{\text{J}}{\text{kg}} = 10^{-2}\,\text{Gy} = 1\,\text{cGy} \tag{B.7}$$

As a result, if an absorbed dose is quoted in rad, it can be expressed directly in cGy. Note that when the symbolic name "rad" might be confused with "rad" for radian (plane angle), it may be shortened to "rd".

B.3 Equivalence between Roentgen and Rad

To find the equivalence between the roentgen and the rad, in order to translate exposure measurements into absorbed dose, the procedure is as described in Chapter 3:

$$1\,\text{R} = 2.58 \cdot 10^{-4}\frac{\text{C}}{\text{kg}} \cdot \frac{10^{-3}\,\text{kg}}{\text{g}} \cdot \frac{33.85\,\text{eV}}{\text{ion pair (air)}}$$

$$\times\;1.6 \cdot 10^{-19}\frac{\text{J}}{\text{ev}} \cdot \frac{1\,\text{ion pair}}{1.6 \cdot 10^{-19}\,\text{C}} \cdot \frac{1\,\text{erg}}{10^{-7}\,\text{J}}$$

$$= 2.58 \cdot \frac{33.85\,\text{ergs}}{\text{g}} = 87\,\text{erg} \cdot \text{g}^{-1} = 0.87\,\text{rad} = 0.87\,\text{cGy} \tag{B.8}$$

Coupled with the radiation weighting factor of 1 for photons, this equation says that an exposure of 1 R to any human person gives him/her an absorbed effective dose of 0.87 rem (approximately 1 rem) when the absorbed dose is measured in units of rad.

As before, it is important to remember that the roentgen is defined only for photons, because it is a unit of exposure. The rad, on the other hand, is defined for all kinds of ionizing radiation.

For photons it is customary to make the following approximation:

$$1\,\text{R} \approx 1\,\text{rad} \approx 1\,\text{rem} \tag{B.9}$$

Therefore:

$$1\,\text{R} \approx 1\,\text{cGy} \tag{B.10}$$

Appendix C

LOGARITHMS

Whenever a number or symbol is raised to a power, such as 2^3, the power is known as the exponent; and the quantity raised is known as the base, or radix. Powers of 10 are quite common in the radiologic and health sciences; 10 constitutes the base.

What are logarithms? The logarithm to the base a of any number b is a number c such that, when a is raised to the power c, it is equal to b:

$$\log_a b = c \tag{C.1}$$

which means that:

$$a^c = b \tag{C.2}$$

As an example, consider the number 1000, which is equal to 10 multiplied by itself three times, or 10 raised to the power of 3. Then:

$$\log_{10} 1000 = 3 \tag{C.3}$$

because:

$$10^3 = 1000 \tag{C.4}$$

When logarithms to the base e (e = 2.7128...) are used, or $\log_e$, it is common to call them natural logarithms and to use the symbol ln. The term "natural logarithms" has come into common usage because many physical laws can be described by equations using base e, and graphs of these quantities are straight lines when the y axis is evaluated in terms of the natural logarithm of the y value, ln y.

The rules for using logs are simple and clear. They are the same as the rules for exponents.

1. When two numbers are multiplied together, add the logs:

$$20 \cdot 4 = 80 \tag{C.5}$$

Using logs:

$$\ln 20 + \ln 4 = \ln 80 \tag{C.6}$$

2. When two numbers are to be divided, subtract the logs:

$$20 / 4 = 5 \qquad \text{(C.7)}$$

Using logs:

$$\ln 20 - \ln 4 = \ln 5 \qquad \text{(C.8)}$$

3. When a number is to be raised to a power, multiply the log by that power:

$$2^5 = 32 \qquad \text{(C.9)}$$

Using logs:

$$5 \ln 2 = \ln 32 \qquad \text{(C.10)}$$

This last equation can also be written as

$$\ln 2^5 = \ln 32 \qquad \text{(C.11)}$$

Hence:

$$5 \ln 2 = \ln 2^5 \qquad \text{(C.12)}$$

4. The log of the base is simply the exponent or power of the base:

$$\ln (e^2) = 2 \qquad \text{(C.13)}$$

5. When the exponent of the base is a log, then the entire quantity is equal to the value of the log:

$$e^{\ln 2} = 2 \qquad \text{(C.14)}$$

6. Anything that has the exponent 0 is equal to 1, so the log of 1 in any base is equal to 0:

$$\ln 1 = 0 \qquad \text{(C.15)}$$

These simple, yet powerful rules are all that is needed to use logarithms successfully. Table C.1 gives the natural logarithms of powers of 10.

Table C.1: Natural logarithms of 10

$\ln 10 = 2.302585$	$\ln 10^4 = 9.210340$	$\ln 10^7 = 16.118096$
$\ln 10^2 = 4.605170$	$\ln 10^5 = 11.512925$	$\ln 10^8 = 18.420681$
$\ln 10^3 = 6.907755$	$\ln 10^6 = 13.815511$	$\ln 10^9 = 20.723266$

Appendix D

GREEK ALPHABET

A	α	alpha
B	β	beta
G	γ	gamma
D	δ	delta
E	ε	epsilon
Z	ζ	zeta
H	η	eta
J	θ	theta
I	ι	iota
K	κ	kappa
L	λ	lambda
M	μ	mu
N	ν	nu
X	ξ	xi
O	o	omicron
P	π	pi
R	ρ	rho
S	σ	sigma
T	τ	tau
U	υ	upsilon
F	φ	phi
X	χ	chi
Y	ψ	psi
W	ω	omega

GLOSSARY

A.

Absorbed Dose: See Dose, Absorbed Dose.

Absorbed Fraction A term used in internal dosimetry. It is the fraction of energy emitted as a specified radiation type i in a specified source region S which is absorbed in a specified target organ or tissue T. (Symbol: $\phi_i(T\leftarrow S)$)

Absorption: The process by which radiation imparts some or all of its energy to any material through which it passes. (See also Compton Effect, Pair Production and Photoelectric Effect.)

    Self-Absorption: Absorption of radiation (emitted by radioactive atoms) by the material in which the atoms are located; in particular the absorption of radiation within a sample being assayed.

Accelerator: A machine that accelerates electrically charged atomic particles to high velocities. Electrons, protons, deuterons, and alpha particles can be accelerated to nearly the speed of light for use in nuclear research. Types of accelerators include the betatron, cyclotron, linear accelerator, and synchrotron.

Activity: The number of nuclear transformations occurring in a given quantity of radioactive material in a unit time. In SI units, activity is measured in s^{-1} and has the special name becquerel (Bq). The previous special unit of activity was curie (Ci). (1 Ci = 37 GBq; 1 Bq = $2.7 \cdot 10^{-11}$ Ci). (Symbol: A or R)

Adsorption: The adhesion of one substance to the surface of another.

Air Kerma: See Kerma.

Algorithm: A formula or set of steps for solving a problem.

Alpha Particle: A charged particle emitted from the nucleus of an atom having a mass and charge equal in magnitude to those of a helium nucleus, i.e., two protons and two neutrons.

Aluminum Equivalent: The thickness of type 1100 aluminum alloy affording the same attenuation, under specified conditions, as the material in question.

Anion: Negatively charged ion.

Annihilation (Electron): An interaction between a positive and a negative electron in which both particles disappear and all the energy, including rest energy, associated with the two particles is converted into electromagnetic radiation (called annihilation radiation).

Anode: This is the (relatively) positive electrode; it is the electrode to which negative ions are attracted.

Aperture (Radiology): The opening in the collimator that allows radiation to reach the detector, e.g., for computed tomography.

Atomic Number: The number of protons in the nucleus of a neutral atom or nuclide. The "effective atomic number" of a compound or mixture is calculated from its composite atomic numbers. The compound or mixture would interact with photons in the same way as an element having this atomic number. (Symbol: Z)

Attenuation (Radiology): The process by which a beam of radiation is reduced in intensity when passing through some material. It is the combination of absorption and scattering processes that leads to a decrease in flux density of the beam when projected through matter.

 Attenuation Coefficient: A general term used to describe quantitatively the reduction in intensity of a beam of radiation as it passes through a particular material.

 Attenuation Coefficient, Compton: The fractional decrease in the intensity of a beam of ionizing radiation due to Compton effect interactions in the medium through which it passes.

 Attenuation Coefficient, Linear: The fractional decrease in the intensity of a beam of ionizing radiation due to all absorption and scattering processes in the medium through which it passes.

 Attenuation Coefficient, Pair Production: That fractional decrease in the intensity of a beam of ionizing radiation due to pair production in the medium through which it passes.

 Attenuation Coefficient, Photoelectric Effect: That fractional decrease in the intensity of a beam of ionizing radiation due to photoelectric effect in the medium through which it passes.

Attenuation Block: A block or stack of material with a cross section larger than the radiation beam with a total thickness equivalent to 3.8 cm of type 1100 aluminum.

Attenuation Factor: A measure of the opacity of a layer of material for radiation traversing it; the ratio of the incident intensity to the transmitted intensity. This ratio is equal to I_0/I, where I_0 and I are the intensities of the incident and emergent radiation, respectively. In the usual sense of exponential absorption, $I = I_0 e^{-\omega t}$, the attenuation factor is $e^{-\omega t}$, where t is the thickness of the material and ω is the absorption coefficient.

Avalanche (Detectors): The multiplicative process in which a single charged particle accelerated by a strong electric field produces additional charged particles through collision with neutral gas molecules. This cumulative increase of ions is also known as "Townsend ionization" or "Townsend avalanche".

Average Life (Mean Life, Radioactive): The average of the individual lives of all the atoms of a particular radioactive substance. It is 1.44 times the radioactive half-life.

Avogadro's Number (Avogadro Constant): Number of atoms in a gram atomic weight of any element; also the number of molecules in a gram molecular weight of any substance. It is numerically equal to $6.023 \cdot 10^{23}$ on the unified mass scale. (Symbol: N_A)

B.

Barriers, Protective: Barriers of radiation-absorbing material, such as lead, concrete, and plaster, used to reduce radiation exposure.

   Barriers, Primary Protective: Barriers sufficient to attenuate the useful beam of an X-ray tube to the required degree.

   Barriers, Secondary Protective: Barriers sufficient to attenuate stray radiation (scattered plus leakage) produced by an X-ray tube to the required degree

Beam: A unidirectional or approximately unidirectional flow of electromagnetic radiation or of particles.

   Useful Beam (Radiology): Radiation that passes through the aperture, cone, or other collimating device of the X-ray source housing; sometimes called "primary beam".

   Beam Hardening: The process of eliminating the low-energy photons from a beam of X-rays. This process changes the quality of the beam in such a manner that the average energy of the beam increases.

   Beam Limiting or Beam Defining Device: A device that provides a means to restrict the dimensions of the useful beam. In any region outside the beam the device, if it is an integral part of the radiation-producing equipment, must have enough shielding to adequately meet the leakage requirement of the source assembly to which it is attached.

Becquerel: The name for the SI unit of activity. One becquerel equals one nuclear disintegration per second (dps). 37 GBq = 1 Ci. (Symbol: Bq)

Bent Beam Linear Accelerator: A linear accelerator geometry in which the accelerated electron beam must change direction by 270° to strike the target.

Beta Particle: Charged particle emitted from the nucleus of an atom, with a mass and charge equal in magnitude to that of the electron.

Biologic Effectiveness of Radiation: See Relative Biologic Effectiveness.

Body Section Radiography: See Tomography.

Bragg-Gray Principle: The relationship between energy absorbed in a small gas-filled cavity in a medium to energy absorbed (in the medium) from ionizing radiation. The relationship is expressed as $E_m = W \cdot J_g \cdot S_g^m$, where E_m equals the energy per mass absorbed in the medium, W equals the average energy needed to produce an ion pair in the gas, J_g equals the number of ion pairs per mass formed in the gas, and S_g^m equals the ratio of the stopping power for secondary particles in the medium to that in the gas.

Branching (Radioactivity): The occurrence of two or more modes by which a radionuclide can undergo radioactive decay. For example, RaC can undergo α or β^- decay, ^{64}Cu can undergo β^-, β^+ or electron capture decay. An individual atom of a nuclide exhibiting branching disintegrates by one mode only. The fraction disintegrating by a particular mode is the "branching fraction" for that mode. The "branching ratio" is the ratio of two specified branching fractions (also called multiple disintegration).

Bremsstrahlung: Secondary photon radiation produced by deceleration of charged particles passing through matter.

Buildup: The increase in absorbed dose with depth below the surface in a material irradiated by a beam of photons or particulate radiation. Buildup may be of two kinds:

   Electron Buildup: This is due to the production by the incident radiation of increasing numbers of forward-moving high-energy electrons increasing with depth until a maximum electron fluence rate has been reached. This effect gives rise to the phenomenon of "skin sparing" and is most marked for photon energies greater than about 400 keV. The effect is not noticeable for X-ray photons generated by potentials of less than 400 kV. For high-energy beams, this process is more important.

   Photon Buildup: Multiple photon scattering in the superficial layers of the phantom, which may lead to an increase in absorbed dose for a short distance. This effect is observed particularly with photons generated by potentials of 50 to 150 kV and large field sizes.

Buildup Factor: The ratio of the intensity of X or gamma radiation (both primary and scattered) at a point in an absorbing medium to the intensity of only the primary radiation. This factor has particular application for "broad beam" attenuation. "Intensity" may refer to energy flux dose, or energy absorption.

Burial Ground (Graveyard): A place for burying unwanted radioactive objects to prevent escape of their radiation, the earth or water acting as a shield. Such objects must be placed in watertight, noncorrodible, containers so the radioactive material cannot leach out and invade underground water supplies.

C.

Calibration: Determination of variation from standard, or accuracy, of a measuring instrument to ascertain necessary correction factors.

Capture, Electron: A mode of radioactive decay involving the capture of an orbital electron by its nucleus. Capture from a particular electron shell is designated as K-electron capture, L-electron capture, etc.

Capture, K-Electron: Electron capture from the K shell by the nucleus of the atom. Also loosely used to designate any orbital electron process.

Capture, Radiative: The process by which a nucleus captures an incident particle and loses its excitation energy immediately by the emission of gamma radiation.

Capture, Resonance: An inelastic nuclear collision occurring when the nucleus exhibits a strong tendency to capture incident particles or photons of particular energies.

Cathode: Negative electrode; electrode to which positive ions are attracted.

Cation: Positively charged ion.

Cells Near Bone Surfaces: Those tissues which lie within 10 μm of endosteal surfaces and bone surfaces lined with epithelium.

Certified Source Assembly: A source assembly certified by an assembler to comply with the leakage requirements of the United States Radiation Control for Health and Safety Act of 1968 (FDA, 1986).

Chamber, Ionization: An instrument designed to measure a quantity of ionizing radiation in terms of the charge of electricity associated with ions produced within a defined volume. (See also Condenser r-Meter.)

  Air-Wall Ionization Chamber: Ionization chamber in which the materials of the wall and electrodes are so selected as to produce ionization essentially equivalent to that in a free-air ionization chamber. This ionization is possible only over limited ranges of photon energies. Such a chamber is more appropriately termed an air-equivalent ionization chamber.

  Extrapolation Ionization Chamber: An ionization chamber with electrodes whose spacing can be adjusted and accurately determined to permit extrapolation of its reading to zero chamber volume.

  Free-Air Ionization chamber: An ionization chamber in which a delimited beam of radiation passes between the electrodes without striking them or other internal parts of the equipment. The electric field is maintained perpendicular to the electrodes in the collecting region. As a result, the ionized volume can be accurately determined from the dimensions of the collecting electrode and the limiting diaphragm. This ionization chamber is the basic standard instrument for X-ray dosimetry within the range of 5 to 1400 kVp.

  Standard Ionization Chamber: A specially constructed ionization chamber from which other ionization chambers can be calibrated.

  Thimble Ionization Chamber: A small cylindrical or spherical ionization chamber, usually with walls of organic material.

  Tissue Equivalent Ionization Chamber: An ionization chamber in which the material of the walls, electrodes, and gas are so selected as to produce ionization essentially equivalent to that characteristic of the tissue under consideration. In some cases it is sufficient to have only tissue equivalent walls, and the gas may be air, provided the air volume is negligible. The essential point in this case is that the contribution to the ionization in the air made by ionizing particles originating in the air is negligible, compared to that produced by ionizing particles characteristic of the wall material.

Chamber, Pocket: A small, pocket-sized ionization chamber used for monitoring radiation exposure of personnel. Before use, it is given a charge, and the amount of discharge is a measure of the radiation exposure.

Charger-Reader: An auxiliary device used for establishing a particular voltage level in an ionization chamber and subsequently for evaluating that voltage level.

Cinefluorography: The production of motion picture photographic records of the image formed on the output phosphor of an image intensifier by the action of X-rays transmitted through the patient (often called cineradiography).

Collimator: A device for confining the elements of a beam within an assigned solid angle. See also beam limiting device.

Collimating Zone: That portion of a diagnostic or therapeutic source assembly that contains the beam limiting device.

Collision: Encounter between two subatomic particles (including photons) which changes the existing momentum and energy conditions. The products of the collision need not be the same as the initial systems.

> **Elastic Collision:** A collision in which no change occurs either in the internal energy of each participating system or in the sum of their kinetic energies of translation.

> **Inelastic Collision:** A collision in which changes occur both in the internal energy of one or more of the colliding systems and in the sums of the kinetic energies of translation before and after the collision.

Committed Effective Dose: See Dose, Committed Effective Dose.

Committed Equivalent Dose: See Dose, Committed Equivalent Dose.

Compton Effect: An attenuation process observed for X or gamma radiation in which an incident photon interacts with an orbital electron of an atom to produce a recoil electron and a scattered photon of energy less than the incident photon. (See also Absorption, Pair Production, and Photoelectric Effect.)

Computed Tomography: An imaging procedure that uses multiple X-ray transmission measurements and a computer program to generate tomographic images of the patients. (Symbol: CT)

Condenser r-Meter: An instrument consisting of an "air-wall" ionization chamber together with auxiliary equipment for charging and measuring its voltage. It is used as an integrating instrument for measuring the air kerma of X or gamma radiation in cGy or R. See also Chamber, Ionization.

Contact Therapy Apparatus: X-ray therapy apparatus designed for very short treatment distance (source to skin distances of 5 cm or less) usually employing peak tube potentials in the range of 20 to 50 kV.

Contamination, Radioactive: Deposition of radioactive material in any place where it is not desired, particularly where its presence may be harmful. The harm may be in vitiating an experiment or a procedure, or in endangering personnel.

Controlled Area: A defined area in which the occupational exposure of personnel (to radiation) is under the supervision of the Radiation Protection Supervisor (also known as the Radiation Protection Officer and the Radiation Safety Officer). This area designation is equivalent to a "restricted area" as defined by the United States Nuclear Regulatory Commission (NRC, 1988).

Conversion Factor: For an image intensifier, this is the quotient of the luminance of the output phosphor divided by the air kerma rate at the input phosphor.

Cortical Bone: That type of bone which is equivalent to "Compact Bone" in ICPR Publication 20 [ICRP, 1973], i.e., any bone with a surface/volume ratio less than 60 cm^2cm^{-3}; in "Reference Man" it has a mass of 4 kg.

Coulomb: Unit of electric charge in the SI system of units. A quantity of charge equal to 1 ampere second.

Count (Radiation Measurements): The external indication of a device designed to enumerate ionizing events. It may refer to a single detected event or to the total number registered in a given period of time. The term often is erroneously used to designate a disintegration, ionizing event, or voltage pulse.

  Spurious Count: In a radiation counting device, a count caused by any agency other than radiation.

Counter, Gas Flow: A device in which an appropriate atmosphere is maintained in the counter tube by allowing a suitable gas to flow slowly through the sensitive volume.

Counter, Geiger-Müller: Highly sensitive, gas-filled radiation-measuring device. It operates at voltages sufficiently high to produce avalanche ionization.

Counter, Proportional: Gas-filled radiation detection device; the pulse produced is proportional to the number of ions formed in the gas by the primary ionizing particle.

Counter, Scintillation: The combination of scintillation, photomultiplier tube, and associated circuits for counting light emissions produced in the phosphors.

Counting, Coincidence: A technique in which particular types of events are distinguished from background events by coincidence circuits, which register coincidences caused by the type of events under consideration.

Counting Ratemeter: An instrument that gives a continuous indication of the average rate of ionizing events.

Cross-Sectional Area (of an X-ray beam): An area in the plane of the beam perpendicular to its direction of travel.

Cross-Talk: This occurs when a source of radioactivity meant to be detected only in one place is actually detected in (an)other place(s) at the same time.

CT Number: One of a set of numbers on a linear scale which are related to the linear attenuation coefficients calculated by a computed tomography device. One of the specific CT numbers on a scale from −1000 for air to +1000 for bone, with water equal to zero, is called a Hounsfield unit.

Curie: The former special unit of activity. One curie equals $3.7 \cdot 10^{10}$ nuclear transformations per second and hence is equivalent to 37 GBq. Several fractions of the curie are in common usage. (Symbol: Ci)

   Microcurie: One-millionth of a curie which represents $3.7 \cdot 10^4$ disintegrations per second and is equivalent to 37 kBq. (Symbol: μCi)

   Millicurie: One-thousandth of a curie which represents $3.7 \cdot 10^7$ disintegrations per second and is equivalent to 37 MBq. (Symbol: mCi)

   Picocurie: One-millionth of a microcurie which represents $3.7 \cdot 10^{-2}$ disintegrations per second or 2.22 disintegrations per minute and is equivalent to 0.037 Bq. (Symbol: pCi; replaces the term μμCi)

Cutie Pie: An ionization chamber device commonly used for detecting radiation exposure rate.

Cyclotron: A particle accelerator in which charged particles receive repeated synchronized accelerations or "kicks" by electrical fields as the particles spiral outward from their source. The particles are kept in the spiral by a powerful magnet.

D.

Daughter: Synonym for Offspring. (See Decay Product.)

Dead Man Switch: A switch so constructed that a circuit-closing contact can be maintained only by continuous pressure on the switch.

Decay, Radioactive: Disintegration of the nucleus of an unstable nuclide by spontaneous emission of charged particles and/or photons.

Decay Constant: The fraction of the number of atoms of a radioactive nuclide which decay in unit time. See also Decay Curve and Disintegration Constant. (Symbol: λ)

Decay Curve: A curve showing the relative amount of radioactive substance remaining after any time interval.

Decay Product: A nuclide resulting from the radioactive disintegration of a radionuclide, formed either directly or as the result of successive transformations in a radioactive series. A decay product may be either radioactive or stable.

Decrement Lines: Imaginary lines drawn through parts where the absorbed energy (radiation dose) is a certain percent of the energy absorbed at the same depth along the central axis of the radiation beam.

Delta Ray: Any secondary ionizing particle ejected by recoil when a primary ionizing particle passes through matter.

Density (Physical): The mass per unit volume of a substance. Usually $kg \cdot m^{-3}$ or $g \cdot cc^{-1}$ (Symbol: ρ)

Detector: An instrument capable of registering the presence of radiation. The two common modes of operation for a detector are:

> **Mean-Level or Integrating:** The average effect of the radiation is accumulated over time.

> **Pulse-Type:** Individual radiation interactions are separated or resolved in time.

Deterministic Effects: Those interactions for which the severity of harm in affected individuals varies with the absorbed dose, and for which a threshold for irreversible biological damage usually exists.

Deuteron: An isotopic form of hydrogen in which the nucleus contains one proton and one neutron. When deuterons are substituted for the common form of hydrogen in the water molecule, the substance is known as "heavy" water.

Diagnostic Source Assembly: A diagnostic source housing (X-ray tube housing) assembly with a beam limiting device attached. This assembly must be constructed so that the leakage radiation air kerma measured at a distance of one meter from the source does not exceed 0.1 cGy in one hour when the source is operated at its leakage technique factors.

Dielectric: A class of material which is generally known as an insulator or non-conductor, but which can support an electric strain. A charge on one part of a non-conductor is not communicated to any other part.

Dielectric Constant: This is a measure of the ease with which a charge carrier can pass through a material. It is known as the relative electrical permittivity.

Dielectric-Relaxation Time: This is the time required for the field due to a charge carrier inside a material such as a semiconductor detector to be communicated to the electrodes. If the dielectric-relaxation time is less than the time required to collect the charge carriers, and if the material in the electrodes is such as to allow charge injection, then secondary currents will flow. The secondary current continues until the charges causing it are either neutralized or collected.

Digital Radiography: A diagnostic procedure using an appropriate radiation source and an imaging system which collects, processes, stores, recalls and presents image information in a digital rather than analog fashion.

Digital Subtraction: An image processing procedure used to improve image contrast by subtracting one digitized image from another.

Directly Ionizing Particles: Charged particles such as alpha or beta particles which cause ionization of an atom without any intermediate interaction taking place.

Disintegration, Nuclear: A spontaneous nuclear transformation (radioactivity) characterized by the emission of energy and/or mass from the nucleus. When numbers of nuclei are involved, the process is characterized by a definite half-life.

Disintegration Constant: The fraction of the number of atoms of a radioactive nuclide which decay in unit time; λ in the equation $N = N_0 e^{-\lambda t}$, in which N_0 is the initial number of atoms present, and N is the number of atoms present after some time t. (See also Decay Constant.)

Dose: A general form denoting the quantity of radiation or energy absorbed. For special purposes it must be appropriately qualified. If unqualified, it refers to absorbed dose. (See also Effective/Equivalent Dose.)

Absorbed Dose: The energy imparted to matter by ionizing radiation per unit mass of irradiated material at a specified point. The SI unit is gray ($1 \text{ J kg}^{-1} = 1$ Gy). The former unit is rad ($1 \text{ rad} = 100 \text{ erg g}^{-1} = 1$ cGy). (Symbol: D. See also Effective/Equivalent Dose.)

Committed Effective Dose: The sum of the products of the committed organ or tissue equivalent doses and the appropriate organ or tissue weighting factors (w_T).

Committed Equivalent Dose: The time integral of the equivalent dose rate in a particular tissue or organ that will be received by an individual following intake of radioactive material into the body. The integration time is 50 years for adults. For children and young persons, doses are calculated to age 70 years.

Cumulative Dose (Radiation): The total dose resulting from repeated exposures to radiation.

Depth Dose: The radiation dose delivered at a particular depth beneath the surface of the body. It is usually expressed as a percentage of surface dose.

Dose Coefficient: The committed tissue equivalent dose per unit intake or committed effective dose per unit intake (Sv Bq^{-1}).

Dose Equivalent: A quantity used in radiation protection. It expresses all radiation on a common scale for calculating the effective absorbed dose. It is defined as the product of the absorbed dose to a tissue or organ multiplied by a quality factor "Q" determined by the properties of the radiation that produced the absorbed dose and certain modifying factors. The SI unit of dose equivalent is the sievert (Sv). The former unit of dose equivalent is rem ($1 \text{ rem} = 1$ cSv). (Symbol: H)

Effective Dose: The sum of the products of the organ or tissue equivalent doses and the appropriate organ or tissue weighting factors (w_T). The SI unit of effective dose is the sievert (Sv). (Symbol: H_E)

Equivalent Dose: The tissue or organ absorbed dose (Gy or J kg^{-1}) multiplied by the radiation weighting factor (w_R). The SI unit of equivalent dose is the sievert (Sv). (Symbol: H_T)

Exit Dose: The radiation dose at the surface of the body opposite to that on which the beam is incident.

Integral Absorbed Dose (Volume Dose): A term used mainly in radiation biology to mean the total energy absorbed by an individual or other biological object or phantom during exposure to radiation. It is frequently obtained by integrating the absorbed dose with respect to mass throughout an irradiated region. It may be stated in joule or kilogram gray.

Maximum Permissible Dose Equivalent: The greatest dose equivalent that a person or specified part thereof shall be allowed to receive in a given period of time. Note that this is presently superceded by the limitations on effective/equivalent dose. (Abbreviated: MPD.)

Median Lethal Dose: Dose of radiation that would be required to kill, within a specified period, 50% of the individuals in a large group of animals or organisms; also called LD_{50}. (Abbreviated: MLD.)

Midline Absorbed Dose: The absorbed dose calculated or measured for a point in tissue and at the "midline" or "center" of the biological specimen, i.e., for the point lying equidistant from the exterior points on the specimen. The designation is for dosimetric purposes and implies no particular biological significance for the midline location.

Percentage Depth Dose: The ratio expressed as a percentage of the absorbed dose rate at a point, at a depth along the beam axis, to the absorbed dose rate at a fixed reference point in the beam axis.

Permissible Dose: The dose of radiation that an individual may receive within a specified period with expectation of no significantly harmful result.

Skin Dose (Radiology): Absorbed dose at center of irradiation field on skin. It is the sum of the dose in air and scatter from body parts.

Threshold Dose: The minimum absorbed dose that will produce a detectable degree of any given effect.

Tissue Dose: Absorbed dose received by tissue in the region of interest expressed in cGy. See also Absorbed Dose.

Dose Meter, Integrating: Ionization chamber and measuring system designed for determining total radiation administered during an exposure. In medical radiology the chamber is usually designed to be placed on the patient's skin. A device may be included to terminate the exposure when it has reached a desired value.

Dose Rate: Absorbed dose delivered per unit time.

Dose Ratemeter: Any instrument that measures radiation dose rate.

Dosimeter: Instrument to detect and measure accumulated radiation exposure. In common usage, a pencil-size ionization chamber with a self-reading electrometer, used for personnel monitoring.

Dosimetry, Photographic: Determination of cumulative radiation dose with photographic film and density measurement.

E.

Edge Enhancement: In xeroradiography, the disproportionate attraction of toner particles toward the region of high residual charge to produce a region of increased perceptibility along an image boundary between two proximal structures.

Efficiency (Counters): A measure of the probability that a count will be recorded when radiation is incident on a detector. Usage varies considerably, so it is well to ascertain which factors (e.g., window transmission, sensitive volume, energy dependence) are included in a given case.

Elective Examination: An examination not requiring immediate execution and therefore able to be planned for the patient's convenience and safety.

Electrode: A conductor used to establish electric contact with a nonmetallic part of a circuit.

Electrometer: Electrostatic instrument for measuring the difference in potential between two points. Used to measure change in electric potential of charged electrodes resulting from ionization produced by radiation.

Electromotive Force: Potential difference across electrodes tending to produce an electric current.

Electron: A stable elementary particle which has the basic unit of electric charge equal to $-1.60210 \cdot 10^{-19}$ C and a rest mass equal to $9.1091 \cdot 10^{-31}$ kg.

   Secondary Electron: An electron ejected from an atom, molecule, or surface as a result of an interaction with a charged particle or photon.

   Valence Electron: Electron that is gained, lost, or shared in a chemical reaction.

Electron Affinity: The tendency of a neutral atom to attract a free electron to itself.

Electron Equilibrium: A condition established in a standard ionization chamber whereby the number of electrons entering a specified volume equals the number of electrons leaving that volume.

Electron Volt: A unit of energy equivalent to the energy gained by an electron in passing through a potential difference of 1 volt. Larger multiple units of the electron volt are frequently used; keV for thousand or kilo electron volts; MeV for million for mega electron volts. (Symbol: eV where $1 \text{ eV} = 1.6 \cdot 10^{-19}$ J)

Electroscope: Instrument for detecting the presence of electric charges by the deflection of charged bodies. It has two metallic leaves hanging at the end of a very slender vane. When like charges are placed on the leaves, they move apart or repel. As the charge is reduced, the leaves move closer together until they are finally side by side when the charge has been reduced to zero.

Element: A category of atoms all having the same atomic number.

Energy: Capacity for doing work. "Potential energy" is the energy inherent in a mass because of its spatial relation to other masses. "Kinetic energy" is the energy possessed by a mass because of its motion; SI units: $kg \cdot m^2 \cdot s^{-2}$ or joules.

    Binding Energy: The energy represented by the difference in mass between the sum of the component parts and the actual mass of the nucleus.

    Excitation Energy: The energy required to change a system from its ground state to an excited state. Each different excited state has a different excitation energy.

    Ionizing Energy: The average energy lost by ionizing radiation in producing an ion pair in a gas or other material.

Energy Dependence: The characteristic response of a radiation detector to a given range of radiation energies or wavelengths compared with the response of a standard free-air chamber.

Energy Fluence: The sum of the energies, excluding rest energies, of all particles passing through a unit cross-sectional area.

Energy Flux Density (Energy Fluence Rate): The sum of the energies, excluding rest energies, of all particles passing through a unit cross-sectional area per unit time (energy fluence per unit of time).

Energy Imparted: See Dose, Integral Absorbed Dose.

Excitation: The addition of energy to a system, thereby transferring it from its ground state to an excited state. Excitation of a nucleus, an atom, or a molecule can result from absorption of photons or from inelastic collisions with other particles.

Exposure: A measure of the ionization produced in air by ionizing radiation. It is the sum of the electric charges on all ions of one sign produced in air when all electrons liberated by photons in a volume element of air are completely stopped in air, divided by the mass of the air in the volume element. The SI unit of exposure is $C \cdot kg^{-1}$ The former unit of exposure is the roentgen ($1 \text{ R} = 2.58 \cdot 10^{-4} \text{ C} \cdot kg^{-1}_{air}$). Note that the concept of exposure has been replaced by that of kerma.

    Acute Exposure: Radiation exposure of short duration.

    Chronic Exposure: Radiation exposure of long duration.

F.

Film Badge: A pack of photographic film that measures radiation exposure for personnel monitoring. The badge may contain two or three films of differing sensitivity and filters to shield parts of the film from certain types of radiation.

Film Ring: A film badge in the form of a finger ring.

Filter (Radiology): Primary: A sheet of material, usually metal, placed in a beam of radiation to absorbed preferentially the less penetrating components. Secondary: A sheet of material of low atomic number (relative to the primary filter) placed in the filtered beam of radiation produced by the primary filter.

Filtration, Inherent (X-Rays): The filter permanently in the useful beam; it includes the window of the X-ray tube and any permanent tube or source enclosure.

> **Added filter** Any filter in addition to the inherent filter.

> **Inherent filter** The filter permanently in the useful beam; it includes the window of the X-ray tube and any permanent enclosure for the tube source.

> **Total filter** The sum of the inherent and added filter.

Fissile (Fissionable) Material: Any material readily fissioned by slow neutrons, for example, ^{235}U, ^{239}P.

> **Fissile Class I Packages:** Those which may by transported in unlimited numbers, in any arrangement, and which require no nuclear criticality safety controls during transportation. However, a transport index based on radiation levels may be required.

> **Fissile Class II Packages:** Those having a transport index of not less than 0.1 and not greater than 10 which may be transported in any arrangement in numbers not to exceed an aggregate transport index of 50, and which require no nuclear criticality control during transportation.

> **Fissile Class III Packages:** Those which do not meet the requirements of Fissile Class I or II packages and which are controlled to provide nuclear criticality control during transportation through arrangements between the shipper and the carrier. Such arrangements must preclude loading, storage, or transport with other fissile material and should provide for: (1) transport in a vehicle assigned for sole use of the consignor with specific instructions and restrictions issued with the shipping papers, or (2) transport under escort by a person in a separate vehicle who has the authority, capability, equipment, and instruction, necessary to provide the administrative controls.

Fission: The splitting of a heavy nucleus into two roughly equal parts (which are nuclei of lighter elements), accompanied by the release of a relatively large amount of energy and frequently one or more neutrons. Fission can occur spontaneously, but usually it is caused by the absorption of gamma-ray photons, neutrons, or other particles.

Fluorography: The production of a photographic record of the image formed on the output phosphor of an image intensifier by the action of X-rays transmitted through the patient.

Focal Spot (X-Rays): The part of the target of the X-ray tube struck by the main electron stream. The "effective" focal spot is the apparent size of the radiation source region in the source assembly when viewed from the central axis of the useful radiation beam.

Fractional Absorption in the Gastrointestinal Tract: The fraction of an element entering the gastrointestinal tract which reaches body fluids is termed the f_i value.

Framing: In cinefluorography, the registration of the circular image of the output phosphor on the rectangular film element or frame.

> **Over-framing:** The entire rectangular frame is filled with the circular image extending beyond the edges of the frame.

> **Under-framing:** The circular image is entirely within the rectangular frame.

Frequency: Number of cycles, revolutions, or vibrations completed in a unit of time. (See also Hertz.)

G.

Gamma-Ray: Short wavelength electromagnetic radiation of nuclear origin (range of energy from 10 keV to 9 MeV), i.e., a photon emitted from the nucleus.

Gas Amplification: As applied to gas ionization radiation detecting instruments, the ratio of the charge collected to the charge produced by the initial ionizing event.

Geiger Region: In an ionization radiation detector, the operating voltage interval in which the charge collected per ionizing event is essentially independent of the number of primary ions produced in the initial ionizing event.

Geiger Threshold: The lowest voltage applied to a counter tube for which the number of pulses produced in the counter tube is essentially the same, regardless of a limited voltage increase.

Geiger Tube: An ionization type radiation detector with a very high sensitivity for photons in the energy range 10 to 1000 keV.

Generator ("Cow"): A device in which a daughter radionuclide is eluted from an ion exchange column containing a parent radionuclide which is long-lived compared to the daughter.

Genetic Effect of Radiation: Inheritable change, chiefly mutations, produced by the absorption of ionizing radiations. On the basis of present knowledge these effects are purely additive; recovery does not occur.

Genetically Significant Dose: That absorbed dose equivalent which, if received by every member of the population, would be expected to produce the same total genetic injury to the population as the actual absorbed dose equivalent received by various individuals.

Geometric Unsharpness: Lack of edge definition present on the recorded image due to the combined optical effect of finite size of the radiation source and geometric separation of the anatomic area of interest from the image receptor and the collimator.

Glove Box: An enclosure used for working with radionuclides particularly those in the form of powders and volatile liquids.

Gray: The special name for the SI unit of absorbed dose, kerma, and specific energy imparted. The gray is equal to one joule per kilogram:$1 \text{ Gy} = 1 \text{ J kg}^{-1} = 100$ rad. (Symbol: Gy).

Ground State: The state of a nucleus, atom, or molecule at its lowest energy. All other states are "excited".

H.

Half-Life: A general term used to describe the time elapsed until some physical quantity has decreased to half of its original value. Here the concept of half-life will be applied to radionuclides.

> **Half-Life, Biologic:** The time required for the body to eliminate 50% of an administered dose of any substance by regular processes of elimination. This is approximately the same for both stable and radioactive isotopes of a particular element in a designated pharmaceutical form.

> **Half-Life, Effective:** Time required for a radioactive element in the body to be diminished by 50% as a result of the combined action of radioactive decay and biologic elimination.

$$\text{Effective half-life} = \frac{\text{Biologic half-life} \cdot \text{Radioactive half-life}}{\text{Biologic half-life} + \text{Radioactive half-life}}$$

> **Half-life, Radioactive:** Time required for a radioactive substance to lose 50% of its activity by radioactive decay. Each radionuclide has a unique half-life.

Half-Value Layer (Half Value Thickness): The thickness of a specified substance which, when introduced into the path of a given beam of radiation, reduces the kerma (exposure) rate by one-half. (Symbol: HVL)

Hardness (X-Rays): A relative specification of the quality of penetrating power of X-rays. In general, the shorter the wavelength the harder the radiation.

Health, Radiologic: The art and science of protecting human beings from injury by radiation, and promoting better health through beneficial applications of radiation.

Heel Effect: Non-uniform intensity observed because a small fraction of the X-ray beam emitted in a direction nearly parallel to the angled target surface must pass through more target material before escaping from the target than does the major portion of the beam which is emitted more perpendicularly. (Note: In addition to the non-uniform intensity the angled target also produces non-uniform image resolution due to variations in apparent focal spot size as viewed from various positions on the image receptor.)

Heredity: Transmission of characters and traits from parent to offspring.

Hertz: Unit of frequency equal to 1 cycle per second. (See also Frequency.)

Hounsfield Units: See CT number.

Hygroscopic: A substance which has the tendency to absorb water, even water vapor from the air is termed hygroscopic. The scintillation material sodium iodide doped with thallium, (NaI(Tl)) is an example of such a material.

I.

Image Intensifier: An X-ray image receptor which increases the brightness of a fluoroscopic image by electronic amplification and image minification.

Image Receptor: A system for deriving a diagnostically usable image from the X-rays transmitted by the patient. Examples: screen-film system; stimulable phosphor; solid state detector.

Image Receptor Assembly: An image receptor in a specialized container necessary for proper operation of the receptor.

Indirectly Ionizing Particles: Particles that cause ionization to occur only after an intermediate interaction producing a charged particle has taken place.

Installation (Radiology): A radiation source with associated equipment, and the space in which it is located.

Intensity: Amount of energy per unit time passing through a unit area perpendicular to the line of propagation at the point in question.

Interlock: A device used to assure proper and safe use of a radiation installation by monitoring (usually by electrical devices) the status, presence or position of various associated devices such as source position, collimator opening, beam direction, door closure, filter presence and preventing the production or emission of radiation if the potential for an unsafe condition is detected.

International Commission on Radiological Protection (ICRP): An international organization, founded in 1928, and supported financially by the World Health Organization (WHO), the International Atomic Energy Agency (IAEA), the United Nations Environment Program, the International Society of Radiology, and others, which operates under rules approved by the International Congress of Radiology. Members of ICRP are selected from nominations submitted to it by the National Delegations to the International Congress of Radiology and by the ICRP itself. The International Executive Committee of the Congress approves the selections.

International Commission on Radiation Units and Measurements (ICRU): An international organization, founded in 1925, with the principle objective of development of internationally acceptable recommendations regarding: (1) quantities and units of radiation and radioactivity, (2) procedures suitable for measurement and application of these quantities in clinical radiology and radiobiology, and (3) physical data needed in the application of these procedures, the use of which, tends to assure uniformity in reporting. The ICRU works closely with the ICRP in its consideration of and recommendations for the radiation protection field. Financially, ICRU is supported by the United States National Institutes of Health and many national and international societies, foundations, and companies.

Inverse Square Law: A rule relating two physical entities by a particular proportionality constant. This constant is one divided by the square of some other physical quantity, e.g., the distance between the two physical entities.

Ion: Atomic particle, atom, or chemical radical bearing an electric charge, either negative or positive.

Ion Pair: Two particles of opposite charge, usually referring to the electron and positive atomic or molecular residue resulting after the interaction of ionizing radiation with the orbital electrons of atoms.

Ionization: The process by which a neutral atom or molecule acquires a positive or negative charge.

> **Primary Ionization:** (1) In collision theory: the ionization produced by the primary particles as contrasted to the "total Ionization" which includes the "secondary ionization" produced by delta rays. (2) In counter tubes: The total ionization produced by incident radiation without gas amplification.

> **Secondary Ionization:** Ionization produced by delta rays.

> **Specific Ionization:** Number of ion pairs per unit path length of ionizing radiation in a medium, e.g., per centimeter of air or per micron of tissue.

> **Total Ionization:** The total electric charge of one sign on the ions produced by radiation in the process of losing its kinetic energy. For a given gas, the total ionization is closely proportional to the initial ionization and is nearly independent of the nature of the ionizing radiation. It is frequently used as a measure of radiation energy.

Ionization Density: Number of ion pairs per unit volume.

Ionization Path (Track): The trail of ion pairs produced by ionizing radiation in its passage through matter.

Ionization Potential: The potential necessary to separate one electron from an atom, resulting in the formation of an ion pair.

Ionizing Event: Any occurrence of a process in which an ion or group of ions is produced.

Irradiation: Exposure to radiation.

Isobars: Nuclides having the same mass number but different atomic numbers.

Isodose Curves (Radiation Oncology): Imaginary lines drawn though a patient or test object along which the radiation absorbed dose is calculated to be some fixed value.

Isomers: Nuclides having the same number of neutrons and protons but capable of existing, for a measurable time, in different quantum states with different energies and radioactive properties. Commonly, the isomer of higher energy decays to one with lower energy by the process of isomeric transition.

Isotones: Nuclides having the same number of neutrons in their nuclei.

Isotopes: Nuclides having the same number of protons in their nuclei, and hence the same atomic number, but differing in the number of neutrons, and therefore in the mass number. Almost identical chemical properties exist between isotopes of a particular element. The term should not be used as a synonym for nuclide.

   Stable Iostope: A nonradioactive isotope of an element.

J.

Joule: The unit for work and energy, equal to one newton expended along a distance of one meter ($1\ J = 1\ N \cdot 1\ m$).

K.

Kerma (Kinetic Energy Released per unit MAss): The kinetic energy of charged ionizing particles liberated per unit mass of specified material by uncharged ionizing particles such as photons and neutrons. Kerma is measured in the same units as absorbed dose, joule per kilogram ($J\ kg^{-1}$) and its special name is gray (Gy). Kerma can be quoted for any specified material at a point in free space or in an absorbing medium. Since air kerma and tissue kerma differ by less than 10% over a wide range of photon energies, these two may be considered equal in magnitude for radiation protection purposes. In this respect, air kerma means air kerma in air. Kerma is independent of the complexities of geometry of the irradiated mass element, and permits, therefore, specification for photons or neutrons in free space or in an absorbing medium and hence has a wider applicability than exposure. (Symbol: K)

Kiloelectronvolt: One thousand electron volts, (10^3 eV). (Symbol: keV)

Kilovolt: A unit of electric potential difference, equal to one thousand volts, (10^3 V). (Symbol: kV)

Kilovolt Constant: The value in kilovolts of the potential difference of a constant potential generator. (Symbol: kVcp)

Kilovolt Peak: The maximum value in kilovolts of the potential difference of a pulsating potential generator. When only half the wave is used, the value refers to the useful half of the cycle. (Symbol: kVp)

L.

Laboratory Monitor: (See Survey Meter.)

Latent Image: The state of developability occurring between the time of exposure of a film to radiation and the processing of that film.

Lead Equivalent: The thickness of lead affording the same attenuation, under specified conditions, as the material in question.

Leakage Radiation: See Radiation.

Leakage Technique Factors: These are specific technique factors (associated with specific source assemblies) which are used in measuring leakage radiation. They are defined as follows:

1. For diagnostic source assemblies (qv)

2. For capacitor energy storage equipment, the maximum rated kV and the maximum rated number of exposures in an hour at the maximum rated kV with the mAs being the greater of 10 mAs or the minimum mAs (allows greatest exposure in an hour) available.

3. For field emission equipment rated for pulsed operation, the maximum rated number of pulses in an hour at the maximum kVp.

4. For all other types of equipment, the maximum rated kVp and the maximum rated continuous tube current for the maximum kVp.

5. For therapeutic source assemblies (qv) for X-ray production at tube potentials below 500 kV, the maximum rated continuous tube current for the maximum kV.

6. For therapeutic source assemblies for X-ray production at tube potentials of 500 kV and above and for gamma source assemblies, see therapeutic source assemblies.

Lesion: A hurt, wound, or local degeneration.

Lepton: A subatomic particle with spin 1/2, which, therefore, obeys fermi statistics. These particles do not experience the strong nuclear force present in the nucleus. The leptons form a family of elementary particles which are light. The lepton family includes the electron, mu-meson, and tau-meson, which are massive and charged, and their associated antiparticles (e.g., the positron). For each charged, massive particle (anti-particle) there is a corresponding neutrino (anti-neutrino) which also has spin 1/2, but neither charge nor rest mass. Some recent, unconfirmed, experiments suggest that the neutino might have mass.

Linear Energy Transfer: The quotient of dE by dL, in which dL is the distance traversed by a particle and dE is the average energy loss in dL due to collisions with energy transfers less than some specific value. Simply, it is a conventional expression for energy deposition measured along the track of an ionizing particle. Gamma and X-ray photons generate low

LET electron tracks. Natural alpha particles and fast neutrons and protons give high-LET tracks. (Symbol: LET)

M.

Magnification (Imaging): An imaging procedure carried out with magnification usually produced by purposeful introduction of distance between the subject and the image receptor.

Mass: The material equivalent of energy; different from weight in that it neither increases nor decreases with gravitational force.

Mass Number: The number of nucleons (protons and neutrons) in the nucleus of an atom. (Symbol: A)

Maximum Permissible Dose Equivalent (MPD): See Dose.

Mean Free Path: The average distance that particles of a specified type travel before a specified type (or types) of interaction(s) in a given medium. The mean free path may thus be specified for all interactions (i.e., total mean free path) or for a particular type of interaction such as scattering, capture, or ionization.

Mean Life: The average lifetime for an atomic or nuclear system in a specified state. For an exponentially decaying system, the average time for the number of atoms or nuclei in a specified state to decrease by a factor of the irrational number e (2.718...).

Megaelectronvolt: One million electron volts, 10^6 eV. (Symbol: MeV)

Milliampere: A unit of current. Generally the current flowing between the filament and anode of an X-ray tube is stated in this unit. (Symbol: mA)

Modulation Transfer Function: A mathematical entity that expresses the relative response of an imaging system or system component to sinusoidal inputs as a function of varying spatial frequency which is often expressed as line pairs per millimeter (lp/mm). The reference value most commonly used is that for zero frequency. This function can be thought of as a measure of spatial resolution of the detector system. (Symbol: MTF)

Molybdenum Breakthrough: This term refers to the amount of parent nuclide, molybdenum, contained in an eluted sample of its offspring ^{99m}Tc. This is allowed to be 4.5 Bq (0.15 μCi) of ^{99}Mo per 37 MBq (1 mCi) of ^{99m}Tc eluate.

Momentum: The product of the mass of a body and its velocity; SI units: $kg \cdot m \cdot s^{-1}$.

Monitoring (Radiologic): Periodic or continuous determination of the amount of ionizing radiation or radioactive contamination present in an occupied region.

Area Monitoring: Routine monitoring of the radiation level of a particular area, building, room, or equipment. Some laboratories or operations distinguish between routine monitoring and survey activities.

Personnel Monitoring: Monitoring any part of an individual, his breath, excretions, or any part of his clothing.

Monoenergetic: Having only one energy associated with it.

Monte Carlo Method: A method permitting the solution by means of a computer of problems in particle physics, such as those of neutron transport, by determining the history of a large number of elementary events and the application to them of the mathematical theory of random variables.

Mutation: Alteration of the usual hereditary pattern.

N.

National Council on Radiation Protection and Measurements (NCRP): This committee was granted a United States Congressional chapter in 1964. It is operated as in independent organization financed by contributions from government, scientific societies, and manufacturing associations.

Natural (Napierian) Logarithms: A system of logarithms using the irrational number e as base.

Negative Ion: Negatively charged ion; commonly termed "anion."

Neoplasm: Any new and abnormal growth, such as a tumor; "neoplastic disease" refers to any disease that forms tumors, whether malignant or benign.

Neutrino: An subatomic particle with spin 1/2, which therefore, obeys fermi statistics. Each neutrino (anti-neutrino) is associated with one of the massive, charged leptons. Neutrinos are uncharged and currently thought to be massless. There has been some recent unconfirmed experimental evidence that suggests neutrinos might have mass.

Neutron: An electrically neutral or uncharged particle of matter existing along with protons in the atoms of all elements except the atomic mass one isotope of hydrogen. The isolated neutron is unstable and decays with a half-life of about 13 minutes into an electron and proton. Neutrons sustain the fission chain reaction in a nuclear reactor.

Epithermal (Slow) Neutron: A neutron having an energy between 0.5 and 10^5 eV. (Sometimes the energy range is given as 0.5 to 100 eV and the energy range from 100 eV to 10^5 eV is called Intermediate.)

Fast Neutron: A neutron having an energy above 0.1 MeV (10^5 eV).

Thermal Neutron: A neutron having an energy of about 0.025 eV which corresponds to a velocity of 200 m/s.

Newton: The unit of force that, when applied to a 1 kilogram mass, will give it an acceleration of 1 meter per second per second ($1 \, N = 1 \, kg \cdot 1 \, m \cdot 1 \, s^{-2}$).

Nomogram: Conversion scale between two sets of units.

Noncontrolled Area: A space not meeting the definition of a controlled area. This area designation is equivalent to an unrestricted area as defined by the United States Nuclear Regulatory Commission [10CRF20, NRC, 1988].

Nonionizing Radiation: Radiation that does not cause ionization when it interacts with matter.

Nonstochastic Effects: See Deterministic Effects.

Nuclear Reaction: See Reaction, Nuclear.

Nuclear Reactor: A device by means of which a fission chain reaction can be initiated, maintained, and controlled. Its essential component is a core with fissionable fuel. It usually has a moderator, a reflector, shielding, and control mechanisms.

Thermal Nuclear Reactor: A nuclear reactor in which the fission chain reaction is sustained primarily by thermal neutrons. Most existing reactors are thermal reactors.

Nucleon: Common name for a constituent particle of the nucleus. Commonly applied to a proton or neutron.

Nucleus (Nuclear): That part of an atom in which the total positive electric charge and most of the mass is concentrated.

Nuclide: A species of atom characterized by the constitution of its nucleus. The nuclear constitution is specified by the number of protons (Z), number of neutrons (N), and energy content; or, alternatively, by the atomic number (Z), mass number A (= N + Z), and atomic mass. To be regarded as a distinct nuclide, the atom must be capable of existing for a measurable time. Thus nuclear isomers are separate nuclides, whereas promptly decaying excited nuclear states and unstable intermediates in nuclear reactions are not so considered.

O.

Occupancy Factor: The factor by which the workload should be multiplied to correct for the degree of occupancy (by any one person) of the area in question while the source is in the "ON" condition and emitting radiation. This multiplication is carried out for radiation protection purposes to determine compliance with the effective dose limits. (Symbol: T)

Oncology: Preferred name for tumor treatment. See Therapy, Radiation Therapy.

Offspring: Synonym for Daughter. See Decay Product.

Operator (Radiation): Any individual who personally utilizes or manipulates a source of radiation.

Optically Stimulated Luminescence: Crystalline materials exposed to ionizing radiation, redistribute the electrical charge within the crystal. In some crystals, a large fraction of displaced charge can become trapped for long periods in higher energy states. The extra energy resulting form the higher electron energy states are released as light photons when crystal is stimulated by certain wavelengths of light. The released light photons are detected using a photomultiplier tube. The signal from the tube can then used to calculate the absorbed dose to the material.

P.

Pair Production: An absorption process for X and gamma radiation in which the incident photon is annihilated in the vicinity of the nucleus of the absorbing atom with subsequent production of an electron and positron pair. This reaction only occurs for incident photon energies exceeding 1.02 MeV. (See also Absorption, Compton Effect, and Photoelectric Effect.)

Parent: A radionuclide that, upon disintegration, yields a specific nuclide—either directly or as a later member of a radioactive series.

Path, Mean Free: Average distance a particle travels between collisions.

Penumbra: The region, at the edge of a radiation beam, over which the absorbed dose rate changes rapidly as a function of distance from the axis. It may be defined geometrically and dosimetrically.

> **Geometric Penumbra:** That region in space which could be irradiated by primary photons or particles coming from part of the source only. By analogy, the transmission penumbra is the region irradiated by photons or particles which have transversed part of the thickness of the collimator, i.e., at its outer edge.

> **Geometric Penumbra Width:** The width of the geometric penumbra in a plane perpendicular to the beam axis at any distance of interest from the source. It is a geometrical concept only and is calculated from the expression

$$W = \frac{c(SSD + d - SCD)}{SCD}$$

> in which c is the source diameter (or effective diameter), SSD + d is the distance from the source to point of interest, and SCD is the distance from the source to the edge of the collimator.

> **Physical Penumbra:** This is a dosimetric concept; the physical penumbra width is the lateral distance between two specified isodose curves at a specified depth.

Periodic Table: An arrangement of chemical elements in order of increasing atomic number. Elements of similar properties are placed one under the other, yielding groups and families of elements. Within each group, a gradation of chemical and physical properties exists but, in general, chemical behavior is similar. From group to group, however, a progressive shift of chemical behavior occurs from one end of the table to the other.

Personnel Monitor: A dosimeter (usually a film badge, thermoluminescent or optically stimulated luminescence device, or ionization chamber) used for determining the exposure to an individual. Such monitoring is required for all persons who are radiation workers.

Phantom: A volume of material approximating as closely as possible the density and effective atomic number of tissue. Ideally a phantom should behave in respect to absorption of radiation in the same manner as tissue. Radiation dose measurements made within or on a phantom provide means of determining the radiation dose within or on the human body under similar exposure conditions. Some materials commonly used in phantoms are water, perspex polystyrene, masonite, pressed wood, and beeswax.

Photodisintegration: Process by which a photon of energy equal to or greater than 7 MeV interacts directly with an atomic nucleus. One of the products which may be produced is a neutron.

Photoelectric Effect: Process by which a photon ejects an electron from an atom. All the energy of the photon is absorbed in ejecting the electron and in imparting kinetic energy to it. See also Absorption, Compton Effect, and Pair Production.

Photon: A quantity of electromagnetic energy (E) whose value in joules is the product of its frequency (ν) in hertz and Planck constant (h). The equation is: $E = h\nu$. See also Radiation.

Pig: A lead-lined container used for storing radionuclides.

Pixel: A two dimensional picture element in the presented image.

Planck Constant: A natural constant of proportionality (h) relating the frequency of a quantum of energy to the total energy of the quantum.

$$h = \frac{E}{\nu} = 6.6256 \cdot 10^{-34} \text{ J} \cdot \text{s}$$

Plateau: As applied to radiation detector chambers, the level portion of the counting rate-voltage curve where changes in operating voltage introduce minimum changes in the counting rate.

Positive Ion: Positively charged ion; commonly called cation.

Positron: Particle equal in mass to the electron and having an equal but positive charge.

Potential Difference: Work required to carry a unit positive charge from one point to another.

Power, Stopping: A measure of the effect of a substance upon the kinetic energy of a charged particle passing through it.

Prompt Excretion: The removal of activity directly from blood to urinary bladder or to the gastrointestinal tract with the clearance half-time assigned to blood.

Proportional Region: Voltage range in which the gas amplification is greater than one, and in which the charge collected is proportional to the charge produced by the initial ionizing event.

Protective Apron: An apron made of radiation absorbing materials, used to reduce radiation exposure.

Protective Barrier: See Barriers.

Protective Glove: A glove made of radiation absorbing materials, used to reduce radiation exposure.

Protective Source Housing: An enclosure, for a gamma-beam therapy source, so constructed that the leakage radiation when the source is in the "ON" position, does not exceed the specified limits (1 cGy/h at 1 m from the source housing when the useful beam is less than 10 Gy/h at 1 m; 0.1% at 1 m of the useful beam otherwise).

Proton: Elementary nuclear particle with a positive electric charge equal numerically to the charge of the electron and with a rest mass of $1.67474 \cdot 10^{-27}$ kg.

Q.

Quality (Radiology): The characteristic spectral-energy distribution of X radiation. It is usually expressed in terms of effective wavelengths of half-value layers of a suitable material; e.g., up to 20 kV, cellophane; 20 to 120 kVp, aluminum; 120 to 400 kVp, copper; over 400 kVp, tin.

Quality Factor (Q): The linear–energy–transfer–dependent factor by which absorbed dose is multiplied to obtain (for radiation protection purposes) a quantity that expresses – on a common scale for all ionizing radiation – the effectiveness of the absorbed dose. This has been replaced by the radiation weighting factor (w_R) and the tissue weighting factor (w_T).

Quantum: An observable quantity is said to be "quantized" when its magnitude is, in some or all of its range, restricted to a discrete set of values. If the magnitude of the quantity is always a multiple of a definite unit, that unit is called the quantum (of the quantity). For example, the quantum or unit of orbital angular momentum is $h/2\pi$, and the quantum of energy of electromagnetic radiation of frequency v is hv. In field theories, a field (or the field equation) is quantized by application of a proper quantum mechanical procedure. This quantization results in the existence of a fundamental field particle, which may be called the field quantum of the electromagnetic field, and in nuclear field theories the meson is considered the quantum of the nuclear field.

Quantum Mottle: The variation in optical density, brightness, CT number or other appropriate parameter in an image which results from the random spatial distribution of the

X-ray or light quanta absorbed at the stage of the imaging chain containing the minimum information content. This stage is known as the quantum sink.

Quantum Theory: The concept that energy is radiated intermittently in units of definite magnitude called quanta, and absorbed in a like manner.

Quenching: The process of inhibiting continuous or multiple discharge in a counter tube which uses gas amplification.

Quenching Vapor: Polyatomic gas used in Geiger-Müller counters to quench or extinguish avalanche ionization.

R.

Rad The former unit of absorbed dose equal to 0.01 joule per kilogram (1 cGy) in any medium. See also Absorbed Dose and Tissue Dose. (Symbol: rad)

Radiation: (1) The emission and propagation of energy through space or through a material medium in the form of waves; for instance, the emission and propagation of electromagnetic waves, or of sound and elastic waves. (2) The energy propagated through space or through a material medium as waves; for example, energy in the form of electromagnetic waves or elastic waves. The term radiation or radiant energy, when unqualified, usually refers to electromagnetic radiation. Such radiation commonly is classified, according to frequency, as hertzian, infrared, visible (light), ultraviolet and X-ray or gamma-ray. (See also Photon.) (3) By extension, corpuscular emissions, such as alpha and beta radiation, or rays of mixed, or of unknown type, as cosmic radiation.

Annihilation Radiation: Photons produced when an electron and a positron unite and cease to exist. The annihilation of a positron–electron pair results in the production of two photons, each having an energy of 0.511 MeV.

Background Radiation: Radiation arising from radioactive material other than the one directly under consideration. Background radiation may also be due to the presence of a radioactive substance in other parts of the building or in the building material itself.

Characteristic (Discrete) Radiation: Radiation originating from an atom after removal of an electron or excitation of the nucleus. The wavelength of the emitted radiation is specific, depending only on the nuclide and particular energy levels involved.

External Radiation: Radiation from a source outside the body—the radiation must penetrate the skin.

Internal Radiation: Radiation from a source within the body (as a result of deposition of radionuclides in body tissues).

Ionizing Radiation: Any electromagnetic or particulate radiation capable of producing ions, directly or indirectly, in its passage through matter.

Leakage (Direct) Radiation: All radiation coming from the X-ray tube source housing except the useful beam.

Monochromatic Radiation: Electromagnetic radiation of a single wavelength, or radiation in which all the photons have the same energy.

Monoenergetic Radiation: Radiation of a given type (e.g., alpha, beta, neutron, gamma) in which all particles or photons originate with and have the same energy.

Primary Radiation: The useful beam of an X-ray tube.

Radiation Weighting Factor: The radiation weighting factor is a dimensionless factor used to derive the equivalent dose from the absorbed dose averaged over a tissue or organ and is based on the quality of radiation. (Symbol: w_R)

Scattered Radiation: Radiation that, during its passage through a substance, has been deviated in direction. It may also have been modified by a decrease in energy.

Secondary Radiation: Radiation resulting from absorption of other radiation in matter. It may be either electromagnetic or particulate.

Stray Radiation: The sum of leakage and scattered radiation.

Useful Beam Radiation: The radiation which passes through the opening in the beam limiting device and which is used for imaging or treatment.

Radiation Protection Survey: An evaluation of the radiation safety in and around an installation, that includes radiation measurements, inspections, evaluations and recommendations.

Radiation Receptor: Any device that absorbs a portion of the incident radiation energy and converts this portion into another form of energy which can be more easily used to produce desired results (e.g., production of an image.) (See image receptor.)

Radiation Source: The region and/or material from which the radiation emanates.

Radioactivity: The property of certain nuclides of (1) spontaneously emitting particles or gamma radiation or (2) emitting X radiation following orbital electron capture or (3) undergoing spontaneous fission.

Artificial Radioactivity: Man-made radioactivity produced by particle bombardment or electromagnetic irradiation, as opposed to natural radioactivity.

Induced Radioactivity: Radioactivity produced in a substance after bombardment with neutrons or other particles. The resulting activity is "natural radioactivity" if formed by nuclear reactions occurring in nature, and "artificial radioactivity" if the reactions are caused by man.

Natural Radioactivity: The property of radioactivity exhibited by more than 50 naturally occurring radionuclides.

Radiography: The making of shadow images on photographic emulsion by the action of ionizing radiation. The image is the result of the differential attenuation of the radiation in its passage through the object being radiographed.

Radiology: That branch of medicine which deals with the diagnostic and therapeutic applications of radiant energy, including X-rays and radionuclides.

Radionuclide: A nuclide that displays the property of radioactivity.

Radiopharmaceutical: A pharmaceutical compound that has been tagged with a radionuclide.

Range: The depth in any material measured from the entrance of an ionizing particle to the stopping position of that particle after it has lost all of its energy.

Reaction (Nuclear): An induced nuclear disintegration, i.e., a process occurring when a nucleus comes in contact with a photon, an elementary particle, or another nucleus. In many cases the reaction can be represented by the symbolic equation: $X + a \rightarrow Y + b$ or, in abbreviated form, $X(a,b)Y$. X is the target nucleus, a is the incident particle or photon, b is an emitted particle or photon, and Y is the product nucleus.

Recombination: The return of an ionized atom or molecule to the neutral state.

Receptor: See Radiation Receptor.

Red Bone Marrow (active): The active red bone marrow is that component of marrow which contains the bulk of the hematopoietic stem cells.

Reference Man: A person with the anatomical and physiological characteristics defined in the report of the ICRP Task Group on Reference Man [ICRP, 1975].

Relative Biological Effectiveness: A factor used to compare the biologic effectiveness of absorbed radiation doses due to different types of ionizing radiation; more specifically, it is the experimentally determined ratio of an absorbed dose of radiation in question to the absorbed dose of a reference radiation required to produce an identical biologic effect in a particular experimental organism or tissue. NOTE: This term should not be used in radiation protection. See also Radiation Weighting Factor. (Symbol: RBE)

Rem: The former special unit of dose equivalent. The dose equivalent is numerically equal to the absorbed dose in rad multiplied by the quality factor, the distribution factor, and any other necessary modifying factors. (Symbol: rem)

Repair: The partial or complete restoration of functional integrity in cells following damage caused by radiation. Operationally, repair means that after irradiation a cell responds as though it had received a smaller dose than under conditions in which damage is more fully expressed. The ability to observe repair implies, therefore, that a comparison is made with a treatment of reference. Full repair indicates that cells respond as though they had not been previously irradiated. (Repair embraces processes sometimes referred to as bypassing of damage, and/or the specific biochemical reversal of damage.)

Resolution: In the context of an image system, the output of which is finally viewed by the eye, it refers to the smallest size or highest spatial frequency of an object of given contrast that is just perceptible. The intrinsic resolution, or resolving power, of an imaging system is measured in line pairs per millimeter (lp/mm), ordinarily using a resolving power target. The resolution actually achieved when imaging lower contrast objects is normally much less, and depends upon many variables such as subject contrast levels and noise of the overall imaging system.

Resolving Time, Counter: The minimum time interval between two distinct events which will permit both to be counted. It may refer to an electronic circuit, a mechanical indicating device, or a counter tube.

Rest Mass: The intrinsic mass of any physical entity; the mass possessed by that entity apart from any motion it may have.

Roentgen: The former special unit of exposure. One roentgen is $2.58 \cdot 10^{-4}$ C kg$^{-1}_{air}$. See also Exposure. (Symbol: R)

Roentgenography: Radiography by means of X-rays.

Rogentgenology: That part of radiology which pertains to X-rays.

Roentgen Rays: X-rays.

S.

Scattered Radiation: See Radiation.

Scattering: Change of direction of subatomic particles or photons as a result of a collision or interaction.

Coherent Scattering: Scattering of photons or particles in which definite phase relationships exist between the incoming and the scattered waves. Coherence manifests itself in the interference between the waves scattered by two or more scattering centers. An example is the Bragg scattering of X-rays and neutrons by the regularly spaced atoms in a crystal, for which constructive interference occurs only at definite angles, called "Bragg angles".

Compton Scattering: The scattering of a photon by an electron. Part of the energy and momentum of the incident photon is transferred to the electron, and the remaining part is carried away by the scattered photon.

Elastic Scattering: Scattering caused by elastic collisions and, therefore, conserving the kinetic energy of the system. Rayleigh scattering is a form of elastic scattering.

Incoherent Scattering: Scattering of photons or particles in which the scattering elements act independently of one another; no definite phase relationships exist among the different parts of the scattered beam. The intensity of the scattered radiation at any point is obtained by adding the intensities of the scattered radiation reaching this point from the independent scattering elements.

Inelastic Scattering: The type of scattering that results in the nucleus being left in an excited state and the total kinetic energy being decreased.

Scattering Coefficient, Compton: That fractional decrease in the energy of a beam of X or gamma radiation in an absorber due to the energy carried off by scattered photons in the Compton effect. See also Compton Absorption Coefficient.

Scintillation Counter: An instrument that detects and measures ionizing radiation by counting the light flashes (scintillations) induced by radiation in certain materials.

Sealed Source: A radioactive source sealed in an impervious container which has sufficient mechanical strength to prevent contact with and dispersion of the radioactive material under the conditions of use and wear for which it was designed.

Secondary Protective Barrier: See Barrier.

Serial Radiography: A radiographic procedure in which a sequence of radiographs is made rapidly by using an automatic cassette changer, image intensifier/TV chain, etc.

Series, Radioactive: A succession of nuclides, each of which transforms by radioactive disintegration into the next until a stable nuclide results. The first member is called the "parent", the intermediate members are called "offsprings" or "daughters", and the final stable member is called the "end product".

Shield: A body of material used to prevent or reduce the passage of particles or radiation. A shield may be designated according to what it is intended to absorb (as a gamma-ray shield or neutron shield), or according to the kind of protection it is intended to give (as a background, biologic, or thermal shield). The shield of a nuclear reactor is a body of material surrounding the reactor to prevent the escape of neutrons and radiation into a protected area, which frequently is the entire space external to the reactor. It may be required for the safety of personnel or to reduce radiation enough to allow use of counting instruments for research or for locating contamination or airborne radioactivity.

Shutter: In beam therapy equipment, a device, attached to the X-ray or gamma-ray source housing to control the "ON" or "OFF" condition of the useful beam.

Sievert: The SI special unit of effective dose, equivalent dose, and dose equivalent. $1 \text{ Sv} = 1 \text{ J kg}^{-1}$ (Symbol: Sv)

SI units: See Appendix A.

Signal to Noise Ratio: The ratio of input signal to background interference. For radiologic images, imaging techniques, etc., the greater the signal to noise ratio, the clearer the image.

Simulator: Diagnostic energy X-ray equipment used to simulate a therapy treatment plan outside the treatment room.

Slice: The single body section imaged in a tomography procedure.

Somatic Effects: Those results of irradiation that may become evident in the individual.

Source: See Radiation Source.

Source Region: That region within the body containing the radionuclide. The region may be an organ, a tissue, or the contents of the gastrointestinal tract, or urinary bladder, or the whole body. (Symbol: S)

Source-to-Image Distance: The distance measured along the central ray from the center of the front of the surface of the source (X-ray focal spot or sealed radioactive source) to the surface of the image detector. (Symbol: SID)

Source-Surface Distance: The distance measured along the central ray from the center of the front surface of the source (X-ray focal spot or sealed radioactive source) to the surface of the irradiated object or person. (Symbol: SSD)

Specific Activity: Total activity of a given nuclide per gram of a compound, element, or radioactive nuclide.

Specific Effective Energy: The energy (MeV), suitably modified for radiation weighting factor, imparted per gram of a target tissue T as a consequence of the emission of a specified radiation (i) from a transformation occurring in source region S. (Symbol: $SEE_i(T \leftarrow S)$)

Specific Gamma-Ray Constant: For a nuclide emitting gamma radiation, the product of exposure (air kerma) rate at a given distance from a point source of that nuclide multiplied by the square of that distance divided by the activity of the source, neglecting attenuation. (Symbol: Γ)

Spectrum: A visual display, a photographic record, or a plot of the distribution of the intensity of radiation of a given kind as a function of its wavelength, energy, frequency, momentum, mass, or any related quantity.

Spot Film: A radiograph taken during a fluoroscopic examination for the purpose of providing a permanent record of an area of interest or to verify the filling of a void with contrast media.

Standard: Something established as a measure; a model to which other similar things should conform.

> **Standard, Radioactive:** A sample of radioactive material, usually with a long half-life, in which the number and type of radioactive atoms at a definite reference time are known. It may be used as a radiation source for calibrating radiation measurement equipment.

Stochastic Effects: Those results of the interaction with radiation the probability of which, rather than the severity, is a function of radiation absorbed dose without threshold. (More generally, stochastic means random in nature.)

Stray Radiation: See Radiation.

Sublethal Damage: Cellular damage, the accumulation of which, may result in lethality.

Survey, Radiologic: Evaluation of the radiation hazards incident to the production, use, or existence of radioactive materials or other sources of radiation under specific conditions.

Such evaluation customarily includes a physical survey of the disposition of materials and equipment, measurements or estimates of the levels of radiation that may be involved, and sufficient knowledge of processes using or affecting these materials to predict hazards resulting from expected or possible changes in materials or equipment.

Survey Meter (Laboratory Monitor): A detection instrument used to monitor an area for unsuspected radiation or to search for a lost radiation source or contamination.

T.

Target (X-ray): The part of an X-ray tube anode assembly impacted by the electron beam to produce the useful X-ray beam.

Target Region: The tissue or organ in which radiation is absorbed. (Symbol: T)

Tenth-Value Layer (Tenth-Value Thickness): The thickness of a specified substance which, when introduced into the path of a given beam of radiation, reduces the kerma rate by ten. (Symbol: TVL)

Therapeutic Source Assembly: A therapeutic source housing assembly for X-ray and electron beam production with a therapeutic beam limiting device attached. There are three types of source assemblies specified and each type has specific leakage limitations.

1. For X-ray production at tube potentials from 5 to 50 kV, the leakage kerma rate at any position 5 cm from the assembly is limited to 0.1 cGy per hour.

2. For X- ray production at tube potentials greater than 50 kV but less than 500 kV, the leakage kerma rate measured at a distance of 1 m from the source in any direction is limited to 1 cGy per hour. Additionally, at a distance of 5 cm from the surface of the assembly, the leakage air kerma is limited to 30 cGy per hour.

3. For X-ray and electron production above 500 kV, the absorbed dose rate due to leakage radiation (excluding that from neutrons) at any point outside the maximum sized useful beam, but within a circular plane of radius 2 m which is perpendicular to and centered on the central axis of the useful beam at the normal treatment distance, is limited to 0.2 percent of the absorbed dose rate to tissue on the central axis at the treatment distance. Furthermore, except for the area defined above, the absorbed dose rate in tissue (excluding that from neutrons) at 1 m from the electron path between the source and the target or the electron window is limited to 0.5 percent of the absorbed dose rate in tissue on the central axis of the beam at the normal treatment distance. Additionally, the contribution of neutrons to the absorbed dose inside the useful beam must be reduced well below 1 percent of the X-ray absorbed dose.

Therapy: Medical treatment of a disease.

Brachytherapy (therapy at short distances): The treatment of disease with sealed radioactive sources placed near, or inserted directly into, the diseased area.

Contact Radiation Therapy: X-ray therapy with specially constructed tubes in which the target–skin distance is very short (less than 2 cm). The voltage is usually 40 to 60 kV.

Radiation Therapy: Treatment of disease with any type of radiation.

Rotation Therapy: Radiation therapy during which either the patient is rotated in front of the source of radiation or the source is revolved around the patient. In this way, a larger dose is built up at the center of rotation within the patient's body than on any area of the skin.

Teletherapy (therapy at long distance): The treatment of disease with gamma radiation at a distance from the patient.

Thermoluminescence: Crystalline materials exposed to ionizing radiation, redistribute the electrical charge within the crystal. In some crystals, a large fraction of displaced charge can become trapped for long periods in higher energy states. This extra trapped energy can be released by heating the material. Some of the released energy appears in the form of light, when the crystal is heated, causing the material to luminesce. This effect is known as thermoluminescence. A photomultiplier tube or light meter can be used to measure the absorbed dose to the material.

Threshold, Photoelectric: The quantum of energy hv_0 that is just enough to release an electron from a given system in the photoelectric effect. The corresponding frequency, v_0, and wavelength, λ, are the threshold frequency and wavelength respectively. For example, in the surface photoelectric effect, the threshold hv_0 for a particular surface is the energy of a photon which, when incident on the surface, causes the electron to emerge with zero kinetic energy.

Tissue Equivalent Material: Material made up of the same elements in the same proportions as they occur in a particular biologic tissue. In some cases, the equivalence may be approximated with sufficient accuracy on the basis of effective atomic number.

Tissue Weighting Factor: The tissue weighting factors are dimensionless factors to derive the effective dose from the equivalent dose. They are based on the difference in sensitivities of tissues to radiation. (Symbol: w_T)

Tomography: A special technique to show in detail images of structures lying in a predetermined plane of tissue while blurring or eliminating detail in images of structures in other planes.

Townsend Avalanche: See Avalanche.

Trabecular Bone: That type of bone which is equivalent to "Cancellous Bone" in ICRP Publication 20 [ICRP, 1973], i.e., any bone with a surface/volume ratio greater than $60 \text{ cm}^2\text{cm}^{-3}$; in Reference Man it has a mass of 1 kg.

Transport Group: This is any one of seven groups into which normal form radionuclides are classified according to their radiotoxicity and potential hazard in transportation.

Transport Index: The number to be placed on a package label to designate the degree of control to be exercised by the carrier during transportation and indicating the following: (1) the highest radiation absorbed dose equivalent rate in microsievert per hour at three feet from any accessible external surface of the package, or (2) for Fissile Class II packages only, the number calculated by dividing the number "50" by the number of similar packages that may be transported together.

U.

Umbra: The region within the beam receiving the full strength of the primary X-ray or gamma-ray photons.

Uncontrolled Area: An area not under the authority of the Radiation Safety Officer and not subject to restriction due to the presence of radiation.

Use Factor (Beam Direction Factor): Fraction of the workload during which the useful beam is directed at the barrier under consideration. (Symbol: U)

Useful Beam: See Radiation.

User: Physicians and others responsible for the radiation exposure of patients.

V.

Valence: Number representing the combining or displacement power of an atom; number of electrons lost, gained, or shared by an atom in a compound; number of hydrogen atoms with which an atom will combine, or which it will displace.

Volt: The unit of electromotive force ($1 \text{ V} = 1 \text{ J·C}^{-1}$).

Voltage: The potential difference, in volts, between two different points in an electric circuit or between two different electrodes.

Voltage, Operating: As applied to radiation detection instruments, the voltage across the electrodes in the detecting chamber required for proper detection of an ionizing event.

Voltage, Starting: For a counter tube, the minimum voltage that must be applied to obtain counts with the particular circuit with which it is associated.

Voxel: A volume element in the object being imaged. In computed X-ray tomography, the mean attenuation coefficient of the voxel determines the CT (Hounsfield) number of the pixel.

Volume, Sensitive: That portion of a counter tube or ionization chamber which responds to a specific radiation.

W.

Watt: The unit of power equal to 1 joule per second $(1 \text{ W} = 1 \text{ J·s}^{-1})$.

Wavelength: Distance between any two similar points of two consecutive waves (λ). For electromagnetic radiation, the wavelength is equal to the velocity of light (c) divided by the frequency of the wave (ν), i.e., $\lambda = c/\nu$.

> **Effective wavelength:** The wavelength of monochromatic X-rays that would undergo the same percentage attenuation in a specified filter as the heterogeneous beam under consideration.

Wave Motion: The transmission of a periodic motion or vibration through a medium or empty space.

> **Transverse:** Wave motion in which the vibration is perpendicular to the direction of propagation.

> **Longitudinal:** Wave motion in which the vibration is parallel to the direction of propagation.

Whole Body Dose Equivalent: The dose equivalent associated with the uniform irradiation of the whole body. (Symbol: H_{wb})

Workload: The degree of use of a radiation source. For X-ray machines operating at tube potentials below 500 kV, the workload is usually expressed in milliampere minutes per week. For gamma—beam therapy sources and for photon—emitting equipment operating at 500 kV or above, the workload is usually stated in terms of the weekly kerma of the useful beam at one meter from the source and expressed in grays per week at one meter. (Symbol: W)

X.

X-rays: Penetrating electromagnetic radiation containing wavelengths shorter than those of visible light. They are usually produced by bombarding a metallic target with fast electrons in a high vacuum. In nuclear reactions, it is customary to refer to photons originating in the nucleus as gamma-rays, and those originating in the extranuclear part of the atom as X-rays. These rays are sometimes called roentgen rays after their discoverer, W.C. Roentgen.

Xeroradiography: The production of an image on a xerographic plate (e.g., electrically charged selenium) by the action of X-rays transmitted through the patient.

> **Xeromammography**: Mammography carried out by the xeroradiographic process.

REFERENCES

Advisory Committee on the Biological Effects of Ionizing Radiation (BEIR I), 1972, *The Effects on Population Exposure to Low Levels of Ionizing Radiation*. Washington, DC, Division of Medical Sciences, National Academy of Sciences, National Research Council

—, (BEIR III), 1980, *The Effects on Population Exposure to Low Levels of Ionizing Radiation*. Washington, DC, Division of Medical Science, National Research Council, National Academy Press.

—, (BEIR IV), 1988, *Health Risks of Radon and Other Internally Deposited Alpha-Emitters*. Washington, DC, Division of Medical Sciences, National Research Council, National Academy Press.

—, (BEIR V), 1990, *Health Risks from Exposure to Low Levels of Ionizing Radiation*. Washington, DC, Division of Medical Science, National Research Council, National Academy Press.

—, (BEIR VI), 1999, *Health Risks of Exposure to Radon*. Washington, DC, Division of Medical Science, National Research Council, National Academy Press.

—, (BEIR VII), 2006, *The Effects on Population Exposure to Low Levels of Ionizing Radiation*. Washington, DC, Division of Medical Science, National Research Council, National Academy Press.

American Association of Physicists in Medicine, 1977, *Basic Quality Control in Diagnostic Radiology*. AAPM Report Number 4. New York, NY, American Institute of Physics.

—, 1993, *Specification and Acceptance Testing of Computer Tomography Scanners*. AAPM Report Number 39. New York, NY, American Institute of Physics.

—, 1980, *Quality Assurance in Diagnostic Radiology*. AAPM Monograph Number 4 New York, NY, American Institute of Physics.

—, 1997, *Code of Practice for Brachytherapy Physics* AAPM Report Number 59. College Park, MD, American Institute of Physics (see also: Medical Physics, 24(10):557-1598, 1997.

—, 2002, *Quality Control in Diagnostic Radiology*. AAPM Report Number 74. College Park, MD, American Institute of Physics.

—, 2006 *PET and PET/CT Shielding Requirements*. AAPM Report Number 108. College Park, MD, American Institute of Physics (see also: Medical Physics, 33(1):4-15, 2006).

Archer, B.R., Thornby, J.I., and Bushong, S.C., 1983, Diagnostic X-ray Shielding Design Based on an Empirical Model of Photon Attenuation. *Health Physics*, **44**:507-517.

Arena, V., 1971, Radiation dose and radiation exposure of the human population. **In:** *Ionizing Radiation and Life*. St. Louis, The C.V. Mosby Co.

Berger, M.J., 1968, Energy Deposition in Water by Photons from Point Isotropic Sources. MIRD Pamphlet 2. *J. Nucl. Med.*, **9**(Suppl.1):15-25.

Bomford, C.K., and Burlin, T.E., 1963, The angular distribution of radiation scattered from a phantom exposed to 100-300 kVp x rays. *Br. J. Radiology*, **36**:436.

Braestrup, C.B., 1944, Industrial Radiation Hazards. *Radiology*, **43**:286.

Bushberg, J.T., Seibert, J.A., Leidholdt Jr., E.M., and Boone, J.M. 2002, *The Essential Physics of Medical Imaging* 2nd Ed. Philadelphia, PA, Lippincott Williams & Wilkins.

Carey, J.E., Kline, R.C., and Keyes Jr., J.W. (eds.), 1983, *CRC Manual Of Nuclear Medicine Procedures*. 4th Ed. Boca Raton, FL, CRC Press.

Code of Federal Regulations, Titles 10 and 49.

Conlon, F.B. and Pettigrew, G.L., 1971, *Summary of Federal Regulations for Packaging and Transportation of Radioactive Materials*. Bureau of Radiological Health, Rockville, MD, U.S. Department of Health, Education and Welfare, BRH/DMRE-71-1.

Cunningham, J.R. and Sontag, M.R., 1980, Displacement corrections used in absorbed dose determination. *Med. Phys.*, **7**:673.

Dillman, L.T., Von der Lage, F.C., 1975, *Radionuclide Decay Schemes and Nuclear Parameters for Use in Radiation Dose Estimation*. MIRD Pamphlet 10. New York, NY, Society of Nuclear Medicine, January.

Dixon, W.R., Garrett, C., and Morrison, A., 1952, Room-protection measures for Cobalt-60 teletherapy units. *Nucleonics*, **10**:42.

Evans, W.W., et al., 1952, Adsorption of 2 MeV constant potential roentgen rays in concrete. *Radiology*, **58**:560.

Hendee, W.R., 1979, *Medical Radiation Physics*. 2nd Ed. Chicago, IL, Year Book Medical
    Publishers.

— and Rossi, R.P., 1979(a), *Quality Assurance for Radiographic X-Ray Units and
    Associated Equipment*. U.S. Department of Health, Education and Welfare, HEW
    Publication (FDA) 79-8094, Bureau of Radiological Health.

— and Rossi, R.P., 1979(b), *Quality Assurance of Fluoroscopic X-Ray Units and
    Associated Equipment*. U.S. Department of Health, Education and Welfare, HEW
    Publication (FDA) 80-8095.

— and Rossi, R.P., 1979(c), *Quality Assurance for Conventional Tomographic X-Ray
    Units*. U.S. Department of Health, Education and Welfare, HEW Publication
    (FDA) 80-8096.

— and Ritenour. E.R., 2002, *Medical Imaging Physics* 4th Ed. New York, John Wiley and
    Sons, Inc.

Institute of Electrical and Electronic Engineers, 1992, *American National Standard Metric
    Practice*. ANSI/IEEE Standard. 268-1992. New York, NY, Institute of Electrical
    and Electronic Engineers, Inc.

—, 1980, *Transactions on Nuclear Science,* Special Issue on Waste Management and
    Storage, **NS-27** Number 4, August.

International Commission on Radiological Protection, 1959, *Recommendations of the
    International Commission on Radiological Protection*. ICRP Publication 2.
    Elmsford, NY, Pergamon Press, Inc.

—, 1968, *Evaluation of Radiation Doses to Body Tissues from Internal Contamination
    Due to Occupational Exposure*. ICRP Publication 10. Elmsford, NY, Pergamon
    Press, Inc.

—, 1966, *Recommendations of the International Commission on Radiological Protection*.
    ICRP Publication 9, Elmsford, NY, Pergamon Press, Inc.

—, 1969, *Radiosensitivity and Spatial Distribution Dose*. ICRP Publication 14. Elmsford,
    NY, Pergamon Press, Inc.

—, 1971, *Assessment of Recurrent or Prolonged Uptakes*. ICRP Publication 10A.
    Elmsford, NY, Pergamon Press, Inc.

—, 1973, *Implications of the Commission's Recommendations that Doses be Kept as Low
    as Readily Achievable*. ICRP Publication 22. Elmsford, NY, Pergamon Press, Inc.

—, 1975, *Reference Man: Anatomical, Physiological and Metabolic Characteristics*.
    ICRP Publication 23. Elmsford, NY, Pergamon Press, Inc.

——, 1977(a), *The Handling, Storage, Use and Disposal of Unsealed Radionuclides in Hospitals and Medical Research Establishments.* ICRP Publication 25. Elmsford, NY, Pergamon Press, Inc.

——, 1977(b), *Recommendations of the International Commission on Radiological Protection.* ICRP Publication 26. Elmsford, NY, Pergamon Press, Inc.

——, 1978, *The Principles and General Procedures for Handling Emergency and Accidental Exposures of Workers.* ICRP Publication 28. Elmsford, NY, Pergamon Press, Inc.

——, 1979(a), *Radionuclide Release Into the Environment: Assessment of Dose to Man.* ICRP Publication 29. Elmsford, NY, Pergamon Press, Inc.

——, 1970(b), *Limits for Intakes of Radionuclides by Workers.* ICRP Publications 30. Elmsford, NY, Pergamon Press, Inc.

——, 1980, *Biological Effects of Inhaled Radionuclides.* ICRP Publication 31. Elmsford, NY, Pergamon Press, Inc.

——, 1982(a), *Protection Against Ionizing Radiation from External Sources Used in Medicine.* ICRP Publication 33. Elmsford, NY, Pergamon Press, Inc.

——, 1982(b), *Protection of the Patient in Diagnostic Radiology.* ICRP Publication 34. Elmsford, NY, Pergamon Press, Inc.

——, 1982(c), *General Principles of Monitoring for Radiation Protection Workers.* ICRP Publication 35. Elmsford, NY, Pergamon Press, Inc.

——, 1989, *Radiological Protection of the Worker in Medicine and Dentistry.* ICRP Publication 57. Elmsford, NY, Pergamon Press, Inc.

——, 1991(a), *Radiation Protection.* ICRP Publication 60. Elmsford, NY, Pergamon Press, Inc.

——, 1991(b), *Annual Limits on Intake of Radionuclides by Workers Based on the 1990 Recommendations.* ICRP Publication 61. (Annals of the ICRP Vol. 21 Number 4) Elmsford, NY, Pergamon Press, Inc.

——, 1996 *Radiological Protection and Safety in Medicine.* ICRP Publication 73. (Annals of the ICRP Vol. 26 Number 2) Elmsford, NY, Pergamon Press, Inc.

——, 1997 *General Principles for the Radiation Protection of Workers.* ICRP Publication 75. (Annals of the ICRP Vol. 27 Number 1) Elmsford, NY, Pergamon Press, Inc.

——, 1997 *Protection from Potential Exposures: Application to Selected Radiation Sources.* ICRP Publication 76. (Annals of the ICRP Vol. 27 Number 2) Elmsford, NY, Pergamon Press, Inc.

—, 1998 *Radiological Protection Policy for the Disposal of Radioactive Waste*. ICRP Publication 77. (Annals of the ICRP Vol. 27 Supplement 1997) Elmsford, NY, Pergamon Press, Inc.

—, 1998 *Individual Monitoring for Internal Exposure of Workers*. ICRP Publication 78. (Annals of the ICRP Vol. 27 Number 3/4) Elmsford, NY, Pergamon Press, Inc.

International Commission on Radiation Units and Measurement, 1970, *Linear Energy Transfer*. ICRU Report Number 16. Washington, DC, ICRU Publications.

—, 1971, *Radiation Protection Instrumentation and its Application*. ICRU Report Number 20. Washington, DC, ICRU Publication.

—, 1972, *Measurement of Low-Level Radioactivity*. ICRU Report Number 22. Washington, DC, ICRU Publications.

—, 1976, International Commission on Radiological Units and Measurement: *Conceptual Basis for the Determination of Dose Equivalent*. ICRU Report Number 25, Washington, DC, ICRU Publications.

—, 1977, *Neutron Dosimetry for Biology and Medicine*. ICRU Report Number 26. Washington, DC, ICRU Publications.

—, 1979, *Methods of Assessment of Absorbed Dose in Clinical Use of Radionuclides*. ICRU Report Number 32. Washington, DC, ICRP Publications.

—, 1980, *Radiation Quantities and Units*. ICRU Report Number 33, Washington, DC, ICRU Publications.

—, 1983, *Microdosimetry*. ICRU Report Number 36. Washington, DC, ICRU Publications.

—, 1984, *Stopping Powers for Electrons and Positrons*. ICRU Report Number 37. Washington, DC, ICRU Publications.

—, 1985, *Determination of Dose Equivalents Resulting from External Radiation Sources*. ICRU Report Number 39, Washington, DC, ICRU Publications.

—, 1992, *Measurements of Dose Equivalent from External Photon and Electron Radiations*. ICRU Report Number 47, Washington, DC, ICRU Publications.

—, 1993(a), *Stopping Powers and Ranges for Protons and Alpha Particles*. ICRU Report Number 49, Washington, DC, ICRU Publications.

—, 1993(b), *Quantities and Units of Radiation Protections*. ICRU Report Number 51, Washington, DC, ICRU Publications.

Johns, H.E. and Cunningham, J.R., 1983, *The Physics of Radiology.* 4th Ed. Springfield, IL, Charles C Thomas.

Karzmark, C.J. and Capone, T., 1968, Measurements of 6 MV x-rays. II Characteristics of secondary radiation. *Br. J. Radiology*, **41**:222.

Kereiakes, J.G. and Rosenstein, M. (eds.), 1980, *CRC Handbook of Radiation Doses in Nuclear Medicine and Diagnostic X-Ray.* Boca Raton, FL, CRC Press, Inc.

Kirn, F.S., Kennedy, R.J., and Wyckoff, H.O., 1954, Attenuation of gamma rays at oblique incidence. *Radiology,* **63**:94.

Knoll, G. F., 1979, *Radiation Detection and Measurement.* New York, NY, John Wiley and Sons, Inc.

Lederer, C.M. and Shirley, V.S., 1978, *Table of Isotopes,* 7th Ed. New York, NY, John Wiley and Sons, Inc.

McGinley, P.H. 1998, *Shielding Techniques for Radiation Oncology Facilities.* Madison, WI, Medical Physics Publishing.

Meredith, W.J. and Massey, J.B., 1977, *Fundamental Physics of Radiology.* 3rd Ed. Bristol, England, John Wright and Sons, Ltd.

Miller, W. and Kennedy, R.J., 1955, X-ray attenuation in lead, aluminum and concrete in the range 275–525 kV. *Radiology*, **65**:920.

National Bureau of Standards, 1977, Guidelines for Use of the Metric System. NBS Letter Circular LC 1056. Washington, D.C., National Bureau of Standards.

——, 1981, The International System of Units (SI). NBS Special Publication 330. Washington, D.C., National Bureau of Standards.

National Council on Radiation Protection and Measurements, 1951(a), *Control and Removal of Radioactive Contamination in Laboratories.* NCRP Report Number 8. Bethesda, MD, NCRP Publications.

——, 1951(b), *Recommendations for Waste Disposal of Phosphorus-32 and Iodine-131 for Medical Users.* NCRP Report Number 9. Bethesda, MD, NCRP Publications.

——, 1953, *Recommendations for the Disposal of Carbon-14 Wastes.* NCRP Report Number 12. Bethesda, MD, NCRP Publications.

——, 1954, *Radioactive Waste Disposal in the Ocean.* NCRP Report Number 16. Bethesda, MD, NCRP Publications.

—, 1959, *Maximum Permissible Body Burdens and Maximum Permissible Concentrations Radionuclides in Air and Water for Occupational Exposure*. NCRP Report Number 22. Bethesda, MD, NCRP Publication.

—, 1961, *Measurement of Absorbed Dose of Neutrons and of Mixtures of Neutrons and Gamma Rays*. NCRP Report Number 25. Bethesda, MD.

—, 1964, *Safe Handling of radioactive Materials*. NCRP Report Number 30. Bethesda, MD, NCRP Publications.

—, 1968, *Medical X-Ray and Gamma-Ray Protection for Energies up to 10 MeV—Equipment Design and use*. NCRP Report Number 33. Bethesda, MD, NCRP Publications.

—, 1971, *Basic Radiation Protection Criteria*. NCRP Report Number 39. Bethesda, MD, NCRP Publications. (Superceded by Report Number 91).

—, 1975(a), *Review of the Current State of Radiation Protection Philosophy*. NCRP Report Number 43. Bethesda, MD, NCRP Publications.

—, 1975(b), *Natural Background Radiation in the United States*. NCRP Report Number 45. Bethesda, MD, NCRP Publications.

—, 1975(c), *Alpha Emitting Particles in the Lungs*. NCRP Report Number 46. Bethesda, MD, NCRP Publications.

—, 1976(a), *Tritium measurement Techniques*. NCRP Report Number 47. Bethesda, MD.

—, 1976(b), *Radiation Protection for Medical and Allied Health Personnel*. NCRP Report Number 48. Bethesda, MD.

—, 1976(c), *Structural Shielding Design and Evaluation for Medical Use of X-Rays and Gamma Rays of Energies up to 10 MeV*. NCRP Report Number 49, Bethesda, MD.

—, 1976(d), *Environmental Radiation Measurements*. NCRP Report Number 50. Bethesda, MD, NCRP Publications.

—, 1977(b), *Cesium-137 from the Environment to Man: Metabolism and Dose*. NCRP Report Number 52. Bethesda, MD, NCRP Publications.

—, 1977(c), *Protection of the Thyroid Gland in the Event of Releases of Radioiodine*. NCRP Report Number 55. Bethesda, MD, NCRP Publications.

—, 1977(d), *Radiation Exposure from Consumer Products and Miscellaneous Sources*. NCRP Report Number 56. Bethesda, MD, NCRP Publications.

—, 1978(a), *Instrumentation and Monitoring Methods for Radiation Protection*. NCRP Report Number 57. Bethesda, MD, NCRP Publications.

—, 1978(b), *A Handbook of Radioactivity Measurements Procedures.* NCRP Report Number 58. Bethesda, MD, NCRP Publications.

—, 1978(c), *Operational Radiation Safety Program.* NCRP Report Number 59. Bethesda, MD, NCRP Publications.

—, 1980(a), *Influence of Dose and Its Distribution in Time on Dose-Response Relationship for Low-LET Radiations.* NCRP Report Number 64. Bethesda, MD, NCRP Publications.

—, 1980(b), *Management of Persons Accidentally Contaminated With Radionuclide.* NCRP Report Number 65. Bethesda, MD, NCRP Publications.

—, 1982, *Nuclear Medicine-Factors Influencing the Choice and Use of Radionuclides in Diagnosis and Therapy.* NCRP Report Number 70. Bethesda, MD, NCRP Publications.

—, 1983(a), *Operational Radiation Safety Training.* NCRP Report Number 71. Bethesda, MD, NCRP Publications.

—, 1983(b), *Radiation Protection and Measurement for Low Voltage Neutron Generators.* NCRP Report Number 72. Bethesda, MD, NCRP Publications.

—, 1983(c), *Protection in Nuclear Medicine and Ultrasound Diagnostic Procedures in Children.* NCRP Report Number 73. Bethesda, MD, NCRP Publications.

—, 1983(d), *I-129-Evaluation of Releases from Nuclear Power Generation.* NCRP Report Number 75. Bethesda, MD, NCRP Publications.

—, 1984(a), *Radiological Assessment: Predicting the Transport, Bioaccumulation, and Uptake by Man of Radionuclides Released to the Environment.* NCRP Report Number 76. Bethesda, MD, NCRP Publications.

—, 1984(b) *Exposure from the Uranium Series with emphasis on Radon and its Daughters.* NCRP Report Number 77. Bethesda, MD, NCRP Publications.

—, 1984(c) *Evaluation of Occupational and Environmental Exposure to Radon and Radon Daughters in the United States.* NCRP Report Number 78. Bethesda, MD, NCRP Publications.

—, 1985, National Council on Radiation Protection: *SI Units in Radiation Protection and Measurements.* NCRP Report Number 82. Bethesda, MD, NCRP Publications.

—, 1987, *Ionizing Radiation Exposure of the Population of the United States.* NCRP Report Number 93. Bethesda, MD, NCRP Publications.

—, 1989, *Medical X-Ray, Electron Beam and Gamma-Ray Protection for Energies up to 50 MeV*. NCRP Report Number 102. Bethesda, MD, NCRP Publications.

—, 1990, *Implementation of the Principle As Low As Reasonably Achievable (ALARA) for Medical and Dental Personal*. NCRP Report Number 107. Bethesda, MD, NCRP Publications.

—, 1992, *Maintaining Radiation Protection Records*. NCRP Report Number 114. Bethesda, MD, NCRP Publications.

—, 1993(a), *Limitation of Exposure to Ionizing Radiation*. NCRP Report Number 116. Bethesda, MD, NCRP Publications. (Supersedes Report Number 91).

—, 1993(b), *Research Need for Radiation Protection*. NCRP Report Number 117. Bethesda, MD, NCRP Publications.

—, 1994, *Dose Control at Nuclear Power Plants*. NCRP Report Number 120. Bethesda, MD, NCRP Publications.

—, 1996, *Sources and Magnitude of Occupational and Public Exposures from Nuclear Medicine Procedures*. NCRP Report Number 124. Bethesda, MD, NCRP Publications.

—, 1997, *Deposition, Retention, and Dosimetry of Inhaled Radioactive Substances*. NCRP Report Number 125. Bethesda, MD, NCRP Publications.

—, 1998, *Radionuclide Exposure of Embryo/Fetus*. NCRP Report Number 128. Bethesda, MD, NCRP Publications.

—, 2001, *Management of Terrorist Events Involving Radioactive Materials*. NCRP Report Number 138. Bethesda, MD, NCRP Publications.

—, 2003(a), *Radiation Protection for Particle Accelerator Facilities*. NCRP Report Number 144. Bethesda, MD, NCRP Publications. (Substantial replaces NCRP Report Number 51.)

—, 2003(b), *Dental X-Ray Protection*. NCRP Report Number 145. Bethesda, MD, NCRP Publications. (Replaces NCRP Report Number 35.)

—, 2004(revised 2005), *Structural Shielding Design for Medical X-Ray Imaging Facilities*. NCRP Report Number 147. Bethesda, MD, NCRP Publications.

—, 2004, *Radiation Protection in Veterinary Medicine*. NCRP Report Number 148. (Replaces NCRP Report Number 36.) Bethesda, MD, NCRP Publications.

—, 2005, *Structural Shielding Design and Evaluation for Megavoltage X- and Gamma-Ray Radiotherapy Facilities*. NCRP Report Number 151. Bethesda, MD, NCRP Publications.

National Research Council, 1976, *A Guide for Preparation of Applications for Medical Programs*. NURGE 0338. Washington, DC, Nuclear Regulatory Commission.

Nationwide Evaluation of X-Ray Trends Task Force (NEXT), 1976, *Suggested Optimum Survey Procedures for Diagnostic X-Ray Equipment*. Co-sponsored by Conference of Radiation Control Program directors and Bureau of Radiological Health (now the Center for Devices and Radiological Health/FDA), Rockville, MD, U.S. Department of Health, Education and Welfare, Bureau of Radiological Health.

Nilsson, B., 1975, Secondary radiation from a spherical tissue-equivalent phantom irradiated with ^{60}Co gamma radiation and 6 MV x rays. *Phys. Med. Biol.*, **20**:963.

Pizzarello, D.J. and Witcofski, R.L., 1975, *Basic Radiation Biology*, 2nd Ed. Philadelphia, PA, Lea & Febiger.

Price,W.J., 1964, *Nuclear Radiation Detection*. 2nd Ed. New York, NY, McGraw-Hill Book Co.

Progress in Radiation Protection, 1975, U.S. Department of Health, Education and Welfare, HEW Publication (FDA) 77-8015. Rockville, MD, Bureau of Radiological Health.

Quality Assurance in Diagnostic Radiology, 1982, Geneva, World Health Organization.

Radiological Health Handbook, 1970, Rockville, MD, Department of Health, Education and Welfare, Public Health Service, U.S. Government Printing Office, Washington, DC, January.

Report of the RBE Committee to the International Commission on Radiation Protection and Radiation Units and Measurements, 1963, *Health Physics*, **9**:357.

Ritz, V.H., 1958, Broad and Narrow beam attenuation of 192 Ir gamma rays in concrete, steel and lead. *Non-destructive Testing*, **16**:269.

Rosenstein, M., (ed.), 1976(a), *Handbook of Selected Organ Doses for Projections Common in Diagnostic Radiology* (Pamphlet). Rockville, MD, U.S.Department of Health, Education and Welfare, Public Health Service, Food and Drug Administration, Bureau of Radiological Health, Number 76-8031, Superintendent of Documents, U.S. Government Printing Office, Washington, DC.

—, 1976(b), *Organ Doses in Diagnostic Radiology*. Rockville, MD, U.S. Department of Health, Education and Welfare, Public Health Service, Food and Drug Administration, Bureau of Radiological Health, Number76-8030, stock Number 017-015-00102-4, Superintendent of Documents, U.S.Government Printing Office, Washington, DC.

Shapiro, J., 2002, *Radiation Protection*. 4rd Ed. Cambridge, MA, Harvard University Press.

Simmons, G.H. and Alexander, G.W., 1970, *Units of radiation exposure and dose.* **In** A Training Manual for Nuclear Medicine Technologists. Rockville, MD, U.S. Department of Health, Education and Welfare, Public Health Service, Bureau of Radiological Health, October.

Simpkin, D.J. 1987, A general solution to the shielding of medical x and c rays by the NCRP report no. 49 methods. *Health Phys.* **52**:431-436.

—,1989, Shielding requirements for constant-potential diagnostic x-ray beams determined by a Monte Carlo calculation x-ray beams determined by a Monte Carlo calculation, *Health Phys.* **56**:151-164.

—,1990, Transmission of Scatter from Computed Tomography (CT) Scanners Determined by a Monte Carlo Calculation. *Health Physics*, **58**:363-367.

—, 1995, Transmission data for shielding diagnostic X-ray facilities. *Health Phys.* **68**:704-709.

—, 1996, Evaluation of NCRP report No. 49 assumptions on workloads and use factors in diagnostic radiology. *Med. Phys.* **23**:577-584.

—, and Dixon RL. 1998, Secondary shielding barriers for diagnostic X-ray facilities: scatter and leakage revisited. *Health Phys.* **74**:350-365.

Snyder, W.S., Fisher Jr., H.L., Ford, M.R., and Warner, G.G., 1969, Estimates of Absorbed Fractions for Monoenergetic Photon Sources Uniformly Distributed in Various Organs of a Heterogeneous Phantom. MIRD Pamphlet 5. *J. Nucl. Med.*, **10**(Suppl.3).

—, Ford, M.R., Warner, G.G., and Watson, S.B., 1975, *"S" Absorbed Dose per Unit Cumulated Activity for Selected radionuclides and Organs.* MIRD Pamphlet 11. New York, NY, Society of Nuclear Medicine, October.

—, Ford, M.R., and Warner, G.G., 1977, *Specific Absorbed Fractions for Radiation Sources Uniformly Distributed in Various Organs of a Heterogeneous Phantom.* MIRD Pamphlet 12. New York, NY, Society of Nuclear Medicine, January.

Spiers, F.W., 1968, *Radioisotopes in the Human Body-Physical and Biological Aspects.* (New York, NY), Academic Press.

Trout, E.D., Kelley, J.P., and Lucas, A. C., 1959, Broad Beam attenuation in concrete for 50–300 kVp x-rays and in lead for 300 kVp x-rays. *Radiology*, **72**:62.

Tsalafoutas,I.A., Yakoumakis, E., and Sandilos,P., 2003, A model for calculating shielding requirements in diagnostic X-ray facilities. *British Journal of Radiology*, **76**:731-737.

United Nations Scientific Committee on the Effects of Atomic Radiation, 1977, *Sources and Effects of Ionizing Radiation*. 1977 Report to the General Assembly. New York, NY, United Nations.

—, 1982, *Ionizing Radiation: Source and Biological Effects*. 1982 Report to the General Assembly. New York, NY, United Nations.

—, 2000, *Sources and Effects of Ionizing Radiation - Vol. 1 Sources, Vol. 2 Effects*. 2000 Report to the General Assembly. New York, NY, United Nations.

—, 2001, *Hereditary Effects of Radiation*. 2001 Report to the General Assembly. New York, NY, United Nations.

Waggener, R.G., Kereiakes, J.G., and Shalek, R.J. (eds.), 1982, *CRC Handbook of Medical Physics*. Boca Raton, FL, CRC Press, Inc.

Wyckoff, H.O., Kennedy, R.J., and Bradford, J.R., 1948, Broad and narrow beam attenuation of 500 to 1400 kVp x-rays in lead and concrete, Radiology, 51:849.

— and Kennedy R.J., 1949, Concrete as a protective barrier for gamma rays from radium. *J. Res.*, NBS, **42**:431.

—, Allisy, A., and Lidén, K., 1976 Communications. *Med. Physics*, **3**:52.

ANSWERS TO PROBLEMS

Chapter 1:

1. 4 keV

2. 0.2410 MeV

3. 1.275 MeV; 1.02 MeV; anti-neutrino has the energy

4. (a) 0.13; (b) 0.82

5. $1.07 \cdot 10^{-4}$ m

6. $7.05 \cdot 10^{-4}$ m

7. (a) 10 keV; (b) no electron released

8. 50 keV

9. 0.49 MeV

10. yttrium; 13, 11, or 10 neutrons are released.

Chapter 2:

1. $1.47 \cdot 10^{4}$ electrons

2. $1.92 \cdot 10^{7}$ electrons

3. A count will be observed

4. (a) 5·103 (b) 12.5%

5. $2 \cdot 10^{6}$ counts

6. (a) $1.67 \cdot 10^{-2}$ s (electron); (b) $0.5 \cdot 10^{-2}$ s (hole)

7. $1.5 \cdot 10^{-3}$ cm

Chapter 3:

1. Both $1 \cdot 10^{-3}$ cGy

2. 0.5 μGy/s

3. $5 \cdot 10^{-4}$ cGy = 5 μGy

4. (a) 1.5 μSv; (b) $1.5 \cdot 10^{-2}$ μSv

5. (a) and (b) 1 cSv

6. (a) 1.01 mSv; (b) $1.2 \cdot 10^{-4}$ Sv

Chapter 4:

1. 50 mSv

2. (a) 380 mSv; (b) 360 mSv

3. (a) 3.6 mSv; (b) 72 mSv; exceeds yearly limit

Chapter 5:

1. Higher energy photons

2. Yes, the alarm will sound ($1.6 \cdot 10^{4}$ cps)

3. 556 Bq

4. $1.02 \cdot 10^{-3}$ mSv

5. Yes, the needle will deflect full scale ($8 \cdot 10^{6}$ cps)

6. (a) 4; (b) The electrometer can measure current; a kVp meter is also needed.

Chapter 6:

1. 2.0 mSv

2. 41.67 mSv/month

3. (a) 22.2 μGy; (b) 22.2 μSv

4. The meter will deflect full scale at an activity level of 3 kBq (3000 cps)

Chapter 7:

1. 556 photons/s

2. (a) $2.5 \cdot 10^5$ photons/s; (b) distance is independent of energy

3. 4 HVL

4. 5 HVL

Chapter 8:

1. No (4.44 kBq)

2. 17 kBq

Chapter 9:

1. $5.4 \cdot 10^2$ L total volume

2. 0.11 cm/s

3. Radioactive yellow II

Chapter 10:

1. 7 HVL

2. 2.8 mm (0.28) cm of lead must be added

3. 0.42 cm of lead must be added

4. 18.5 cm of lead must be added

Chapter 11:

1. (a) 2.67 days; (b) 8 days

2. $9.5 \cdot 10^{-2}$ cGy

3. (a) $5.4 \cdot 10^{-3}$ cGy; (b) $4.32 \cdot 10^{-3}$ cGy (c) 2.16 h

4. $1.47 \cdot 10^{-3}$ cGy

Chapter 12:

1. The f factors are:

	0.02 MeV	2.0 MeV
water	34.3	37.2
muscle	35.8	37.6
bone	115	40

2. 1.5 mGy

3. $1.66 \cdot 10^{-7}$ C

4. Following the structure of Example 12.2:

	CASE A	CASE B
View, projection	AP shoulder, left and right shoulders on separately exposed radiographs	AP shoulder; both shoulders on one radiograph
The active bone marrow absorbed dose for 0.5 cGy $(0.571 \cdot 0.8.73$ cGy) is:	$7.41 \cdot 10^{-5}$ Gy (for standard man where the shoulders are assumed to be 31 cm from the phantom vertex)	$11.99 \cdot 10^{-5}$ Gy

INDEX

A

absorbed dose 59, 61–63, 65–68, 71–73, 75–77, 82, 97, 140–142, 145, 148, 150, 153–154, 157, 181, 193, 202, 205, 208, 211, 213, 223–226, 229–230, 232–233, 239–240, 245, 248, 253–255, 260, 262–263, 268, 270–273, 277–279

absorbent paper 158, 161

absorption 65, 66, 154, 156–157, 162, 245–246, 248, 250, 257–259, 268–269, 272, 275

accelerator 73, 148, 167–168, 245, 247,252

acceptable limits 3, 78

acceptor-rich 49–50

accident 69, 72–73, 75–77, 95, 162

accidental 79, 91, 143, 153

accidentally 71, 77, 117, 160

accountability 170

accumulated activity 211, 213, 219

ACR See American College of Radiology

acute 77, 80

adult human phantom 208

AEC See Atomic Energy Commission

aerosols 118, 154

aggregate 40

agreement cities 175

agreement states 103, 175

air kerma 62, 85, 100, 103, 107, 137–138, 143, 149–151, 157, 181–182, 187, 224, 226, 245

air kerma rate 100, 142, 145, 147, 182–184, 198–199

airborne materials 154

airborne radiation 118

airborne radioactivity 108, 118, 162, 275

airborne radioactivity area 173

air-equivalent material 87, 98

ALARA See As Low As Reasonably Achievable

ALI See Annual Limits of Intake

allowable electronic states 42

allowable internal body 159

allowable limit 181, 188

allowable transmission 192

alpha decay 8, 10

alpha emitter 153, 156

alpha particle(s) 8, 12, 14–16, 21, 23, 32, 40, 43, 48, 57, 64, 86, 88–90, 92, 126, 150, 152–154, 245, 253, 265

alpha radiation(s) 87, 271

alpha-emitting radionuclides 71, 150

alumina 172

American Association of Physicists in Medicine 103, 141, 186, 195, 197–198

American College of Radiology 141, 169

AMU see atomic mass units

angiographic 140–141

angiography 100, 108, 113, 144, 233

animal carcasses 177, 179

animal excreta 177

animal experiment 170, 171

anion 3

annealed 111–112

annihilation photons 19, 102, 182, 185, 186

Annual Limit of Intake 79, 154

Annual Reference Levels of Intake 79,80

anode 30–39, 46, 57–58

anodes 39, 41

anthracene 42

antiscatter grid 145

apparati 141

Apron 270

apron 146, 152

argon 31–32, 37, 40, 58, 95

ARLI see Annual Reference Levels of Intake 79

As Low As Reasonably Achievable 74, 78, 117, 122, 138, 181–182, 185

atmospheric pressure 29, 37

atomic bomb victims 71

SSD See source-skin distance
standard for length 236
sterility 76
stochastic effects 65–67, 69, 74, 78–79, 276
stopping-power 228
storage area 150, 159, 163, 174–175, 178
surface-barrier 50–52, 56
surface-junction 52
survey monitor 85–86, 88, 91–92, 102
Sv See sievert

T

target organ 205–206, 208–209, 211,
    219–220, 245
temperature dependence 89, 91
tenth-value layer 133, 277
thermoluminescent 111
thermoluminescent crystal 111
thermoluminescent devices 27, 40, 230,
    269
thermoluminescent dosimeter(s) 111–112,
    115
tissue kerma 62, 149, 263
tomography 45, 278
tongs 151, 156, 160
Townsend avalanche 33, 36, 246
trabecular bone 153, 209, 278
transportation 147, 167, 175, 258, 278–279
TVL See tenth-value layer

U

Ukraine 69, 72
ultrasonic emissions 169
ultraviolet photon(s) 18, 41, 271
ultraviolet radiation 3, 45, 112
umbra 201, 279
uncontrolled area 181–182, 195, 279
unit of absorbed dose 61, 236
unit of exposure 59, 239–240, 257
unit of kerma 62, 223
United Nations Scientific Committee on

the Effects of Atomic Radiation 2,
    72–73, 75
United Nations World Health Organization
    167
United States Center for Devices and
    Radiological Health/FDA 230
United States Congress 73, 167–169
United States Food and Drug
    Administration 2, 169–170, 233
United States National Institute for Science
    and Technology 102–103
United States National Academy of
    Sciences 4
United States National Council on
    Radiation Protection and
    Measurements 2, 14, 63–65, 67,
    72–80, 144, 148, 182, 188, 191,
    198–199, 266
United States National Institutes of Health
    262
United States National Research Council 4,
    71
United States Nuclear Regulatory
    Commission 73, 78, 103, 158,
    169–171, 175, 178, 251
United States Postal Service 175
United States Radiation Control for Health
    and Safety Act of 1968 168, 249
United States Department of Health and
    Human Services 169
Unites States National Research Council 73
UNSCEAR See United Nations Scientific
    Committee on the Effects of
    Atomic Radiation
unsealed source(s) 147–149 181–182, 198,
    205